Ancient DNA

Bernd Herrmann   Susanne Hummel
Editors

Ancient DNA

*Recovery and Analysis of Genetic Material
from Paleontological, Archaeological, Museum,
Medical, and Forensic Specimens*

With 50 Illustrations

Springer-Verlag

New York   Berlin   Heidelberg   London   Paris
Tokyo   Hong Kong   Barcelona   Budapest

Bernd Herrmann
Institut für Anthropologie
Universität Göttingen
37073 Göttingen, Germany

Susanne Hummel
Institut für Anthropologie
Universität Göttingen
37073 Götttingen, Germany

Cover art provided by Prof. Dr. Bernd Herrmann.

Library of Congress Cataloging-in-Publication Data.
Ancient DNA / [edited by] Bernd Herrmann, Susanne Hummel.
    p.   cm.
    Includes bibliographical references and index.
    ISBN 0-387-97929-8.—ISBN 3-540-97929-8
    1. DNA—Analysis.   2. Paleobiology.   3. Paleopathology.
  I. Herrmann, Bernd. 1946–   . II. Hummel, Susanne.
    [DNLM: 1. DNA.   2. Fossils.   3. Paleontology.   QU 58 A541]
  QP620.A53 1993
  574.87'3282—dc20
  DNLM/DLC   92-49549
  for Library of Congress

Printed on acid-free paper.

Acquiring editor: Robert C. Garber
Production managed by Theresa Kornak, manufacturing supervised by Jacqui Ashri
Typeset by Thomson Press India Ltd., New Delhi
Printed and bound by Braun-Brumfield, Inc., Ann Arbor, MI
Printed in the United States of America

9 8 7 6 5 4 3 2 1

ISBN 0-387-97929-8 Springer-Verlag New York Berlin Heidelberg
ISBN 3-540-97929-8 Springer-Verlag Berlin Heidelberg New York

Preface

When we began our attempts to clone DNA from archaeological specimens of bones in 1986 we were not successful, either with DNA or with funding. We assume that other scholars had similar experiences. Since that time, the field of ancient DNA recovery and analysis has developed into one of the most exciting approaches to studying the past. Application of polymerase chain reaction (PCR) technology to ancient DNA has revolutionized the field within the past few years. Interest in recovering DNA from unconventional material, particularly old specimens, has ranged from the purely scientific to very practical applications, such as law enforcement and medicine.

We are pleased that many of the "first generation" scholars involved in ancient DNA research have contributed to this book. It is our intention to present a general overview of the field before the diversity of disciplines and scientific questions separates workers too much from central issues and themes. We express our thanks to the contributors and also to the organizations that provided funds to support this research in the days before the remarkable successes that are described in this book were demonstrated. Their faith in "high risk" enterprises is vital to the health of scholarly endeavor.

Bernd Herrmann
Susanne Hummel
Göttingen, Germany
31 May 1992

Contents

Contributors

Augusto Amici
Dipartimento di Biologia, Universitá degli Studi di Camerino, I-62032 Camerino, Italy

Raul J. Cano
Biological Sciences Department, California Polytechnic State University, San Luis Obispo CA 93407, USA

Alan Cooper
Molecular Genetics Laboratory, National Zoological Park, Washington, DC 20008, USA

Cynthia D. Dickel
Departments of Immunology and Medical Microbiology, University of Florida, Gainesville, FL 32610, USA

Hans Ellegren
Department of Animal Breeding and Genetics, Swedish University of Agricultural Sciences, S-75007 Uppsala, Sweden

Jan Engberg
The Royal Danish School of Pharmacy, Department of Biological Sciences, DK-2100 Copenhagen, Denmark

Jörg T. Epplen
Molekulare Humangenetik, Gebäude MA, 5.0G Nord, Ruhr-Universität Bochum, D-4630 Bochum, Germany

Edward M. Golenberg
Department of Biological Sciences, Wayne State University, Detroit, MI 48202, USA

WAYNE W. GRODY
Division of Medical Genetics and Molecular Pathology, and Diagnostic Molecular Pathology Laboratory, Department of Pathology and Laboratory Medicine, UCLA School of Medicine, Los Angeles, California 90024-1732, USA

ERIKA HAGELBERG
MRC Molecular Haemotology Unit, Institute of Molecular Medicine, University of Oxford, John Radcliffe Hospital, Oxford OX3 9DU, UK

WILLIAM W. HAUSWIRTH
Departments of Immunology and Medical Microbiology, University of Florida, Gainesville, FL 32610, USA

BERND HERRMANN
Institut für Anthropologie, Universität Göttingen, 37073 Göttingen, Germany

SUSANNE HUMMEL
Institut für Anthropologie, Universität Göttingen, 37073 Göttingen, Germany

DAVID A. LAWLOR
Department of Immunology, University of Texas, M.D. Anderson Cancer Center, Houston, TX 77030, USA

HENRIK NIELSEN
Department of Biochemistry B, The Panum Institute, University of Copenhagen, DK-2200 Copenhagen N, Denmark

GEORGE O. POINAR JR.
Department of Entomological Science, University of California – Berkeley, Berkeley, CA 94720, USA

HENDRIK N. POINAR
Biological Sciences Department, California Polytechnic State University, San Luis Obispo, CA 93407, USA

PETER K. ROGAN
The Milton S. Hershey Medical Center, Department of Pediatrics, Pennsylvania State University, Hershey, 17033, USA

FRANCO ROLLO
Dipartimento di Biologia Molecolare, Cellulare Animale, University of Camerino, 1-62032 Camerino, Italy

JOSEPH J. SALVO
Environmental Research Center, General Electric Co., Research &
Development Center, Schenectady, NY 12301-0008, USA

GEORGE F. SENSABAUGH
Forensic Science Group, Department of Biomedical and Environmental
Health Sciences, School of Public Health, University of California – Berkeley,
Berkeley, CA 94720, USA

ERIC C. SWANN
Department of Plant Biology, University of California – Berkeley, Berkeley,
CA 94720, USA

JOHN W. TAYLOR
Department of Plant Biology, University of California – Berkeley, Berkeley,
CA 94720, USA

INGOLF THUESEN
The Carsten Niebuhr Institute, University of Copenhagen, DK-2300
Copenhagen, Denmark

FRANCO VENANZI
Dipartimento di Biologia, Universita degli Studi di Camerino, I-62032
Camerino, Italy

FRANCIS X. VILLABLANCA
Department of Integrative Biology and Museum of Vertebrate Zoology,
University of California, Berkeley, CA 94720, USA

1 Introduction

BERND HERRMANN and SUSANNE HUMMEL

The detection of high molecular organic compounds in ancient remains has turned out to open a new research area with many implications, the most important being the extraction of ancient DNA (aDNA) (to be characterized below) from fossils, subfossil remains, artifacts, traces from biological sources, and museum specimens. Since this technique provides access to the basic molecules of organismic evolution, it opens up the possibility of studying evolution at the molecular level over a principally unlimited time scale.

It should not be forgotten that knowledge on the preservation of nucleic structures and nucleic acids has been available for a surprisingly long time. As early as the beginning of our century, scholars of plant biology tried to sprout seeds from excavations after they had succeeded in demonstrating nucleic materials by staining old plant tissues. Although not purposely, similar evidence of nucleic materials was obtained in microscopic (histo-chemical) inspections by the pioneers of human paleopathology investigating ancient mummified remains.

Successful extraction of aDNA and aRNA was carried out, presumably for the first time, by a Chinese team from rib cartilage of the Old Lady of Mawangtui, a corpse preserved for almost 2,000 years (Hunan Medical College 1980). The introduction of molecular cloning techniques initiated a breakthrough in the search for aDNA. The first successful amplification of aDNA from animal tissues, reported by Higuchi et al. (1984), was soon followed by molecular cloning of aDNA from Egyptian mummies (Pääbo 1985). Unfortunately, this technique required so much aDNA to be extracted from the source that it was (and is) restricted to extraordinarily well-preserved samples.

However, the invention of the *Polymerase Chain Reaction* (PCR) in the mid-1980s (Saiki et al. 1985; Mullis and Fallona 1987) led to a boom in the search for aDNA and aRNA, since it required only traces of target nucleic acid. Thus it became the method of choice for all researchers interested in molecular approaches to the past. In fact, the past in this context includes all eras of biological evolution up to the present, since it turns out that the basic problems in handling DNA traces from recent biological material are

1

of the same kind as those involving DNA traces from organisms that lived millions of years ago. So the term "ancient DNA" (aDNA) covers any bulk or trace of DNA from a dead organism or parts of it, as well as extracorporeally encountered DNA of a living organism. Therefore, any DNA that has undergone autolytic or diagenetic processes or any kind of fixation is considered to be "aDNA". Access to aDNA provides opportunity to study genetic material of past organisms for individual features and evolutionary processes. Three major issues are of interest:

1. *Access to genetic information at the individual level*: This is the first objective of the analysis of biological evidence material and environmental monitoring as well as of research on patient tissue samples stored in surgical pathology or autopsy files and on plant and animal samples from museum collections. Typing the individual's genetic profile, or that of an organism contained in these samples, is the first step required for all additional research. It is also a prerequisite of

2. *Access to genetic information at the infrapopulation level*: Comparing genetic information from two or more individuals means to determine their relationship in terms of biological distance (i.e., kinship). The similarity or diversity among individuals of a population plays a basic role in evolutionary processes and leads to

3. *Access to genetic information at the interpopulation level*: Comparison of profile of different populations reveals their degree of evolutionary relationship. It also offers the opportunity to reconstruct population history in time and space thus to reconstruct evolutionary trees.

This book reflects all three approaches of DNA analysis. It opens in Chapter I with an overview on aspects of kinship both at the infrapopulation and the interpopulation level. The second section of this volume, Sample Preparation and Analysis, offers basic as well as more sophisticated knowledge and advice for the handling and processing of aDNA from various sources. As it turns out, the problems of extraction and analysis of aDNA are basically the same, regardless of source. Differences are more likely due to the conditions of preservation, e.g., relative humidity or type of fixative used.

The book was compiled to serve both as textbook and as "cookbook", covering most of the important contributions to the field; it is intended to provide guidance in research design.

Since in principle, any preserved organism or parts of it (or products made from biological materials) are suppliers of aDNA, the number of sources is immense. However, various kinds of preservation, degradation, and diagenetic influence may alter the original tissue or remains. Thus, the problems of manipulating different kinds of RNA or DNA from living organisms are comparatively minor compared to the problems of handling and processing aDNA.

Metazoan tissues and plant tissues certainly differ from each other in their prospects of preservation. The fixative used for preservation influences the content of aDNA in a source material (cf. Chapter Taylor and Swann; Post 1992). Compared to parenchymatic tissues, hard tissues such as bones or seeds tend to preserve better, since they lack enzymes and water and offer mechanical protection. There are more reports on successful amplifications of ancient mitochondrial DNA (mtDNA)* than on nuclear DNA**. This reflects the advantage of working with mtDNA, both for evolutionary (cf. Chapter 3, Villablanca) and technical reasons: highly repetitive sequences of mtDNA are abundant in each cell. Not much knowledge is available on aRNA; so far, reports on successful extractions are scarce (cf. Chapter Rollo et al.). Diagenetic factors have more destructive effects on this single-stranded molecule. The fact is that research on the availability and successful amplification of aDNA and aRNA is still mainly following the principle of trial and error. The benefit of the PCR technique is also its limitation: The search for a certain target sequence does not provide simultaneous information on the presence of any other aDNA or aRNA sequences. An improvement of screening methods for total aDNA and aRNA content in sources is needed. Most recently, a generalized PCR method has been reported that will amplify fragments of aDNA without any sequence information using a single primer (Foo et al. 1992). If the sequences found can be identified, this will certainly constitute a remarkable advance.

Much work must also be done to learn how to cope with aDNA damage. Radiation (mainly UV), temperature, moisture, pH, oxidative agents, and mechanical stress are among the most important factors influencing the survival of aDNA under diagenesis. Their impact on aDNA is not yet understood in detail, as research has not focussed extensively on that subject up to now. Only a few contributions have been made so far (cf. Chapter 17, Golenberg; Eglington and Logan 1991). The problem of damage is most obvious with respect to base alterations found in aDNA. A sequence of aDNA may differ from the expected sequence for three reasons: because of an original mutation, diagenetic damage, or an amplification error.

Usually an original mutation can be distinguished from diagenetic sequence artifacts or amplification errors by direct sequencing of the PCR products. In the case of an original mutation, every sequence will reveal the exchanged base. In all other cases, only a few templates or products, respectively, will carry the erroneous base; its signal is weak compared to the signal of the majority of nonaffected sequences. However, basic difficulties of technical approach and interpretation remain if the reaction contains only very few intact template sequences and/or if a nonrandom loss or exchange of sequences

---

*(cf. Chapter 13 Hagelberg)
**(cf. Chapter 14 Hummel and Herrmann)

or single bases in the aDNA must be presumed, for instance the depurination of DNA in an acid environment. First steps toward repair of damaged aDNA have been made (cf. Chapter 8, Nielsen et al.), but again, this field is still waiting to be explored.

There are various ways to successfully extract DNA from ancient source material, as the chapters of this book illustrate. It has become obvious that sample pretreatment and extraction procedure must take into account contamination by either modern DNA of diverse origin and/or ancient microbial DNA. In the beginning, any presence of DNA in ancient materials was attributed to modern contaminants by some, but these doubts were comparatively easy to dispel when proper laboratory work excluded such contamination. Moreover, the introduction of adequate independent control samples (cf. Chapter 4) allows us today not only to confirm the true sample origin of investigated DNA but also to ensure that amplification is not erroneously done from microorganismal sequences.

It is a prerequisite for any successful extraction of aDNA that the material studied contains extractable amounts of ancient nucleic acids, but no general criteria exist for estimating DNA content of ancient specimens before attempting extraction. Developing such criteria would certainly require much research in the difficult area of taphonomy and decomposition, whereas so far most researchers have been interested first and foremost in the fact itself that aDNA could be extracted and amplified. As work on aDNA proceeded, it became clear that almost any biological material from the past may still contain aDNA. (While this book is being prepared, there are indications that aDNA can be traced back at least to the age of the dinosaurs.) Contaminating DNA of micro- and macroorganisms which invaded the sample, or came in contact with it, can certainly cover up persisting endogenous DNA. Therefore, screening methods to estimate the ratio of endogenous to contaminating DNA would be advantageous.

However, it is still advisable simply to test for aDNA instead of using screening methods. Conceivable methods are either not simple, or not specific for aDNA unless they amount to full characterization of the aDNA sequences (cf. Chapter 4). Future developments will certainly improve the situation by allowing in situ hybridization in cross-sections or on surfaces of ancient source material. An aDNA sequence of interest might be found through labelled probes, preferably under light fluorescent microscopy control. Thus PCR would become unnecessary at least in some areas of aDNA research.

Some general remarks on the quality of source material may be in order here. First, as clear from the discussion above, aDNA research is done on damaged, decomposed, and fixated materials. For artificial preservation, the conditions and consequences of fixation are quite well understood from work in light and electron microscopy (e.g., Robinson et al. 1987). Unfortunately, this does not hold true for damaged, decomposed, and fixated samples from "natural environments". Although decomposition research has increased along with the interest in trace elements, stable isotopes, and macromolecules from ancient materials (Boddington et al. 1987; Schwarcz et al. 1989; Grupe and Garland 1992; Lambert and Grupe 1993), much more effort is needed to gain knowledge in "molecular paleontology". This effort must concentrate on the degradation and decomposition of macromolecules, to the understanding of which Eglington and Logan (1991) made one of the important first contributions.

As mentioned above, the standard techniques available today for inspecting ancient tissues and materials are also helpful in decision making and setting up research designs. The most trivial but nonetheless important observation is that there seems to be a general tendency for the quality of preservation to decrease with time after death. However, due to environmental conditions at the locations where materials are found, there are so many exceptions to this rule that one should not take it too seriously. Looking at samples at the microscopic or ultrastructural level yields a lot of information on the materials investigated. To get a rough idea of the quality of preservation and the ratio of aDNA to contaminating DNA, we strongly recommend a microscopic approach to all aDNA work, as illustrated by Figures 1–8.

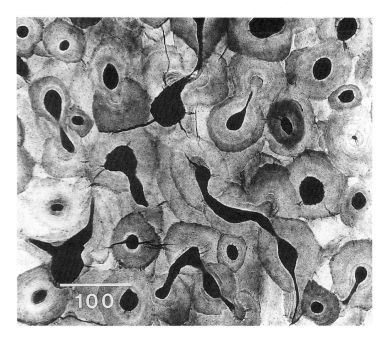

FIGURE 1. Microradiograph of a cross-section of a mammoth femur from the Neanderthal site Salzgitter, Lower Saxony, Germany. Although some 60,000 years old, the microstructure is still perfectly preserved due to wet and anaerobic conditions in glacial gravels and sands. Magnification bar in µm.

FIGURE 2. Close-up view of a fossil magnolia leaf in Pliocene marl sediment. Tne extraordinarily good conditions for preservation allowed persistence of major structural components of the plant soft tissue. Probing for aDNA should be promising in fossils of this quality. Willershausen, Germany, No. 52-12698 of the collection of the Department of Geology and Palaeontology, University of Göttingen. Magnification bar in cm.

Undoubtedly, work on aDNA has serious ethical implications. These do not so much refer to law enforcement, or epidemiological or therapeutic applications. If it turns out that analyzing aDNA from pathological samples furthers our knowledge of infectious diseases and thus improves our strategies in fighting epidemics, such work should be carried out, we feel. Similarly, if the aDNA sequence of an early cultivated plant would benefit extant strains of that species, for instance by improving microbial resistance or crop yield, one would hardly hesitate to introduce the ancient sequence into the contemporary genome, in an effort to stave off nutritional problems for the world population by working against genetic erosion.

But these scenarios nonetheless describe the introduction of foreign DNA into living eukaryotic cells, the creation of new life-forms in vitro. (The insertion of foreign sequences into phages and bacteria in cloning techniques is considered a difficult matter, at least for the most part). Many scientists

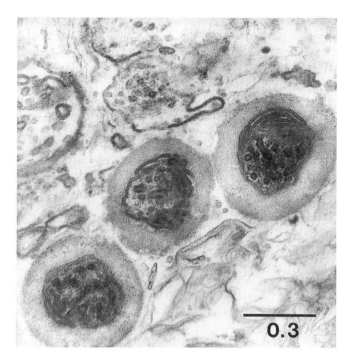

FIGURE 3. TEM picture of remarkably well preserved intracellular ultrastructures, most likely mitochondria, from retrobulbal adjacent tissues of the eyeball of a wet/ anaerobically preserved medieval burial. (From Piepenbrink and Herrmann 1988.) Such structures obviously indicate presence of aDNA. Magnification bar in μm.

believe this step to be ethically questionable. Even if such experiments are done with the best of intentions, problems will obviously arise when sequences of, say, "Lucy" or *Homo erectus*, William Shakespeare or Czarina Catherine become available.

There may not be a generally applicable ethical position on this. The intention and purpose of the experiments as well as the competence and skill of the scientists involved are decisive for the ethical acceptability of work with aDNA. Both the scientific community and the public should be aware of future problems, since work with aDNA will become a pretty easy procedure eventually. Although at present reality lags far behind the "Jurassic Park" vision, any steps toward the realization of such vision would be prevented, since among other issues the natural rights of a chimera thus created will imply unsolvable problems and conflicts with other organisms and ecological systems, both philosophically and biologically.

Ethical problems are less likely to result from mere access to genetic information about deceased individuals. We do not expect problems in collecting information from human skeletal series consisting of hundreds of

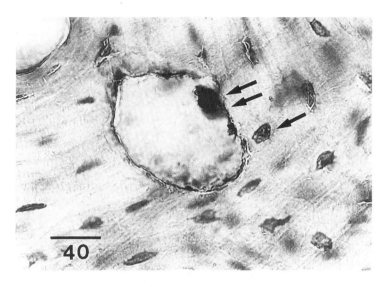

FIGURE 4. While soft tissues covering the skeleton decompose, vestiges may remain in hollow spaces of the hard tissue. Standard methods of soft tissue staining often give positive results, indicating the presence of aDNA. Shown here is the villanueva stain of a cross-section of a femur from a medieval burial in ordinary sandy soil, exhibiting remains of osteocytes (✓) and osteoclasts (✓ ✓). Magnification bar in μm.

anonymous individuals; however, questions may arise as soon as individuals known by name are concerned. Although there is no basic difference between a publication on syphilitic signs on the skeleton of Ulrich von Hutten (d. 1523) and a discussion of President Lincoln's genetic disorders, in the latter case the personal rights of living descendants may be infringed on. To scholars willing to work on subjects like these we recommend contacting local ethics commissions. They also ought to reflect on the epistemological aspects of their projects: what will be gained by the sought-after knowledge, and what might this knowledge be used for?

This book owes its existence to the rare event of a true scientific break-through: the invention of the PCR technique. We are convinced that work with aDNA will soon become routine in the vast majority of biological disciplines, including those using bioproducts and artifacts. Applications in any of these disciplines have been included in this book. One very important application in environmental monitoring, the tracing of a pollutant back to its cause, is missing, however, this dynamic area of research and development is just arising as the book is being produced.

The objectives of future research on aDNA, and the related scientific problems, are made clear by the contributions to this book. The future of

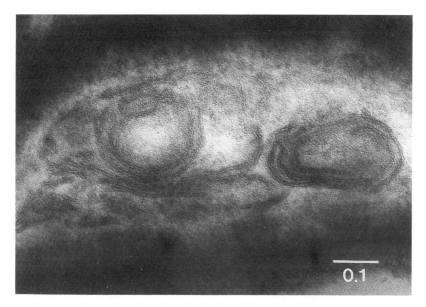

FIGURE 5. Even if standard staining techniques or histochemical tests do not give positive results, degrading cell remnants are still very likely present. This TEM picture shows degrading membranes, indicating a good chance for extracting aDNA, from an ultrathin section of a human femur bone from an eighteenth-century Spitsbergen whaler cemetery. The membranes stem from tissue remains adhering to the walls of a Volkmann's canal. Magnification bar in μm. (Photo: J. Kerl, Department of Anthropology, University of Göttingen.)

this challenging field depends mostly on (1) the improvement of extraction procedures from source materials, including automatic extraction (cf. Chapter 14, Hummel and Herrmann); (2) increased knowledge of degrading processes which affect DNA during its diagenesis, since current knowledge is insufficient for a comprehensive and systematic research program; (3) improved screening for total aDNA and aRNA content, the ability to extract and amplify DNA sequences of more than several hundred base pairs, and the development of repair techniques to restore sequences of appropriate length. We hope this book will prove an important and useful step toward these aims.

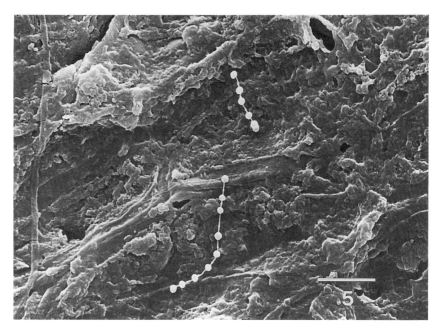

FIGURE 6. Prospects for isolating endogenous DNA can be limited in the presence of contaminating DNA from invading microorganisms, as can be seen in this SEM picture of a 1,200-year-old *Rubus* sp. seed from a Lombard site in northern Italy. (From Rollo et al. 1987.) Magnification bar in μm.

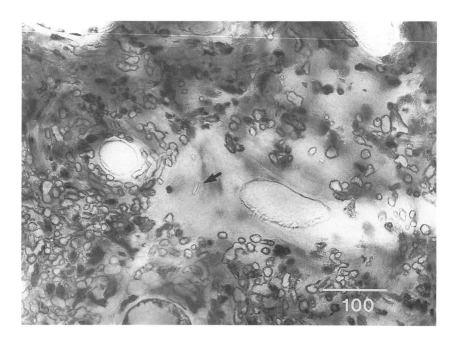

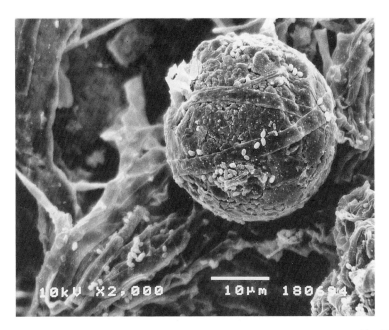

FIGURE 8. Remarkable contamination of a sample is demonstrated by this SEM picture of a decomposing medieval human bone, exhibiting a persisting stage ("cyst") of an unknown plant or lower animal. Covered with actinomycetal hyphes it provides also a base for bacteria. Thus at least four species would contribute to aDNA extraction despite possible contaminations of operators and researchers on that sample. (Courtesy of Karin Rathje, Dept of Anthropology, University of Göttingen). Magnification bar in μm.

◄ ───

FIGURE 7. Microorganisms provide the majority of contaminating DNA under decomposition of any source material. Thus the ratio of aDNA to contaminating DNA may be very low as in this example. Numerous boring canals by invading bacteria and fungi have destroyed the original texture of a human femur from the Romano-British cemetary at Poundbury, England, leaving only a small island of well preserved tissue, enclosing an osteocyte marked by 1. Magnification bar in μm.

References

Boddington A, Garland AN, Janaway RC (eds) (1987) *Death, Decay and Reconstruction: Approaches to Archeology and Forensic Science.* Manchester: Manchester University Press

Eglington G, Logan GA (1991) Molecular preservation. Phil Trans R Soc Lond B 333:315–328

Foo I, Salo WL, Aufderheide AC (1992) PCR libraries of ancient DNA using a generalized PCR method. BioTechniques 12:811–814

Grupe G, Garland AN (eds) (1992) *Palaeohistology: Methods and Diagnosis.* Heidelberg: Springer-Verlag

Higuchi R, Bowman B, Freiberger M, Ryder OA, Wilson AC (1984) DNA sequences from a quagga, an extinct member of the horse family. Nature 312:282–284

Hunan Medical College (1980) Study of an ancient cadaver in Mawangtui Tomb No. 1 of the Han Dynasty in Changsha. Beijing: Ancient Memorial Press, (184–187)

Lambert J, Grupe G (eds) (1993) Prehistoric Human Bone Archaelogy at the Molecular Level. Heidelberg: Springer-Verlag

Mullis KB, Faloona FA (1987) Specific synthesis of DNA in vitro via a polymerase-catalysed chain reaction. Meth Enzymol 155:335–350

Pääbo S (1985) Molecular cloning of ancient Egyptian mummy DNA. Nature 314:644–645

Post RJ (1992) Methods for the preservation of small insects for DNA studies. aDNA Newsletter 1:24–25

Robinson DG, Ehlers U, Herken R, Herrmann B, Mayer F, Schürmann FW (1987) Mehods of Preparation for Electron Microscopy. Heidelberg: Springer-Verlag

Rollo R, La Marca A, Amici A (1987) Nucleic acids in mummified plant seeds: screening of twelve specimens by gel-electrophoresis, molecular hybridization and DNA cloning. Theor Appl Genet 73:501–505

Saiki RK, Scharf S, Faloona F, Mullis KB, Horn GT, Erlich HA, Arnheim N (1985) Enzymatic amplification of β-globulin genomic sequences and restriction site analysis for diagnosis of sickle cell anemia. Science 230:1350–1354

Schwarcz HP, Hedges REM, Ivanovich M (eds) (1989) First International Workshop on Fossil Bone. Appl Geochem 4(3)

Access to Kinship and Evolution

DNA Fingerprinting
2 Simple Repeat Loci as Tools for Genetic Identification

Jörg T. Epplen

1. Introduction

The genomes of eukaryotic organisms harbor two principally different classes of DNA: Single-copy DNA is present once per haploid chromosome set, whereas repetitive DNA can exist in a few to a few hundred thousand (or million) copies. The whole array of repetitive sequences may comprise from less than 10% to more than 90% of the genome in different animal or plant species (in man it is about 30%). Novel names have been given to several subclasses of repetitive DNA elements: *SINES* and *LINES* represent supergroups of *short* and *long interspersed nucleotide elements,* rather loosely defined on the basis of length (for a general review see, e.g., Epplen 1988; Singer 1982). Some designations for more narrowly defined groups reflect their discoverers' whim. After the sputniks had been successfully launched and Yuri Gagarin came back from space, the term *satellite DNA* was coined to designate the sizeable amount (up to 30%) of a eukaryotic genome that can be separated by buoyant density gradient centrifugation. For the present discussion, *minisatellites* and *microsatellites* are of special interest: the expression "minisatellites" was established in order to distinguish a particular kind of tandemly reiterated repeat with a basic unit of roughly 33 base pairs from all others (Jeffreys et al. 1985); later, the designation "microsatellites" caught on, somewhat unfortunately, for tandemly organized stretches of dinucleotides which are amplified by the polymerase chain reaction (PCR). *Variable number of tandem repeats (VNTRs)* is commonly used to name both entities of sequences.

The oldest and more general term *simple repetitive DNA* has then been redefined to refer to short (less than 10 bases long) sequence motifs that are tandemly reiterated many times in a perfect manner (Skinner 1977; Tautz and Renz 1984). While such elements are missing in most prokaryotes, specific sets of simple repetitive sequences are one of the few common items all eukaryotes investigated share in their present-day DNA (for review see Epplen et al. 1991). The overall organization of these simple repetitive elements differs unpredictably, even between closely related members of animal and plant families. In the species studied most intensively to date (man and mouse), the distribution of simple repeats does not appear to be completely random, and higher representations in certain regions of the chromosomes have been discovered, e.g., by in situ hybridization experiments (see below; Nanda et al. 1991). Many of these repeat elements were identified inadvertently in the close vicinity of coding exons, as pure by-products of genomic sequence analyses (Weber 1990). Considering all the information accumulated, it appears reasonable to conclude that the inherent properties of the eukaryotic genetic machinery create such elements and allow them to persist in one form or another, while those of most prokaryotes do not.

One major purpose of this chapter is to discuss critically the practical meaning of simple repeats as diagnostic research tools. Previously, multilocus DNA fingerprinting represented the leading technology in this respect. Thanks to the advent of enzymatic DNA amplification via PCR, a wealth of additional approaches is now feasible. In general, investigating small, known single-copy parts of genomes of old specimens of an existing species has not posed truly insurmountable problems, except that the respective loci were not present in intact, cohesive form. However, so far there are very few reports on experiences with the amplification of simple repeats from ancient DNA (see, e.g., Hagelberg et al. 1991b; Roewer et al. 1991). We cannot presume, of course, that all eukaryotic organisms present on earth in former times were characterized by ubiquitous simple repetitive constituents. If so, students of ancient DNA (aDNA) would be in an excellent position, with efficient tools in place for unraveling some of the secrets of the evolutionary past. Here, we would merely like to provide a solid basis for the discussion of the *applications* of simple repetitive sequences in various methodological approaches, setting aside the unsolved mysteries of their biology.

2. Methodology of Multilocus Fingerprints

In this section, we will summarize the methodology of multilocus fingerprints and the identification of hypervariable single loci by probing and amplification via PCR. A thorough account of the methodological development and advantages of multilocus fingerprinting has been given previously (Epplen 1992).

2.1 DNA Extraction

DNA is isolated from nucleated cells of solid organs or peripheral blood (leukocytes) following established protocols that result in DNA cleavable by commercially available restriction enzymes. Needless to say, in many instances skipping the phenol/chloroform extraction step requires more laborious purification procedures subsequently. "Quick prep" procedures (e.g., Genomix, Kontron) appear quite promising. For plant DNA particular modifications and precautions are necessary. For PCR destruction of the nuclei and boiling of the sample is usually sufficient.

2.2 DNA Digestion

DNA ($3–10\,\mu g$ per human individual; in other species as well, 10^6 haploid genome equivalents are the preferred amount for rapid signal generation) is digested with restriction enzymes according to the manufacturers' recommendations. In general, restriction enzymes with four-base-pair recognition sites (*Hin*fI, *Mbo*I, *Alu*I, *Hae*III) are used for DNA fingerprinting purposes. A digest using a conventional restriction enzyme (that recognizes a sequence of four bases) produces on average about ten million DNA fragments per vertebrate haploid genome. The use of the most informative restriction enzyme *Hin*fI requires particular precautions to ensure complete digestion.

2.3 DNA Electrophoresis

In order to be able to differentiate between particular sequences, given the overwhelming abundance of genetic material in higher eukaryotes, the DNA fragments have to be ordered according to length by electrophoretic separation. Two alternative buffer systems are used extensively: TAE and TBE. TAE offers some advantages for the separation of high molecular weight fragments ($> 10–15\,kb$), whereas TBE, unlike TAE, does not require recirculation or exchange due to exhaustion in longer runs (TAE: $40\,mM$ Tris-HCl, pH 8.3, $12\,mM$ sodium acetate, $2\,mM$ EDTA, pH 8.3; TBE: Tris $10.8\,g/l$, boric acid $5.5\,g/l$, EDTA $0.95\,g/l$). Electrophoresis is usually performed in a 0.7% agarose gel to resolve 1.5-kb- to 30-kb-long DNA fragments. Other agarose concentrations may be advantageous for different fragment length ranges. Depending on the type and size of the electrophoresis chamber, running conditions will vary (up to 48 hr at $1–2\,V/cm$). Each gel slot is filled with the same amount of DNA in the same loading volume ($3–10\,\mu g$). Phenol/chloroform extraction (and subsequent ethanol precipitation) prior to loading is not mandatory, though it may well improve the resolution of the multilocus pattern due to less variation in protein and salt concentration.

2.4 Southern Blotting and Hybridization with Digoxigenated Probes

For hybridization and signal detection, restriction-enzyme-digested and size-fractionated genomic DNA is transferred onto immobilizing membranes. Depurination and base cleavage of DNA, facilitating the transfer of large fragments, is done by treatment with $0.25\,M$ HCl. After denaturation, DNA is blotted under alkaline conditions onto polyvinylide difluoride (PVDF; Biorad or Millipore) membranes, which yield the optimal signal-to-background ratios. Nylon may create variable background problems, whereas nitrocellulose appears less advantageous. Allow hybridization to proceed for about 3 hr at the respective temperature:

– 40°C for dig.-$(GTG)_5/(CAC)_5$
– 38°C for dig.-$(GACA)_4$, dig.-$(CA)_8/(GT)_8$, dig.-$(GGAT)_4$, dig.-$(GAA)_6$
– 30°C for dig.-$(GATA)_4$

according to Zischler et al. 1989). The hybridization solutions may be reused many times for several months.

Digoxigenated oligonucleotides are detected with a monospecific antibody coupled to alkaline phosphatase (Boehringer-Mannheim). If the antibody solution is applied in rolling cylinders, it is important to avoid covering parts of the membrane by overlapping. The antibody solution can also be spotted directly onto the membrane. The phosphatase staining reaction is performed at pH 9.5 in the presence of Mg^{2+} ions. Dye substrates must not be allowed to precipitate.

2.5 Multilocus Probing with ^{32}P-labeled Oligonucleotides

^{32}P from $[\gamma\text{-}^{32}P]ATP$ is transferred to the 5′-end of the oligonucleotide. This reaction is catalyzed by the enzyme T4 polynucleotide kinase. Since only one ^{32}P atom can be incorporated into each oligonucleotide molecule by the T4 kinase, the efficiency of the labeling reaction and the resulting specific activity of the probe are limiting, i.e., these parameters determine the exposure time. The oligonucleotide is separated from unincorporated $[\gamma\text{-}^{32}P]ATP$ by means of ion-exchange chromatography. Chromatography on Bio-gel P_4 or Sephadex columns yields less satisfactory results. Note: This method does not separate labeled from unlabeled oligonucleotides. To increase the specific activity of the labeled probe, purification via denaturing polyacrylamide (PAA) gel is recommended. In principle, thin layer chromatography can also be performed, depending on the equipment available.

Hybridization is allowed to proceed for 1–3 hr (or overnight, depending on one's schedule); then washes are performed according to Schäfer et al. (1988) at $Td - 5°C$ (Melting point of given probe $-5°C$; melting point "calculation" 2°C per A/T base, 4°C per G/C base):

– 45°C for $(GTG)_5/(CAC)_5$, $(TCC)_5$
– 43°C for $(GACA)_4$, $(GGAT)_4$, $(GAA)_6$, $(CA)_8/(GT)_8$
– 35°C for $(GATA)_4/(CTAT)_4$

These solutions may be used several times; they are stored at 4°C or at $-20°C$, depending on anticipated storage time. Sharper bands are obtained without intensifying screens. As a rough guideline, exposure is two to three times longer without screen than with one screen. Dried gels are especially suitable for repeated rehybridizations with different probes. The above oligonucleotide

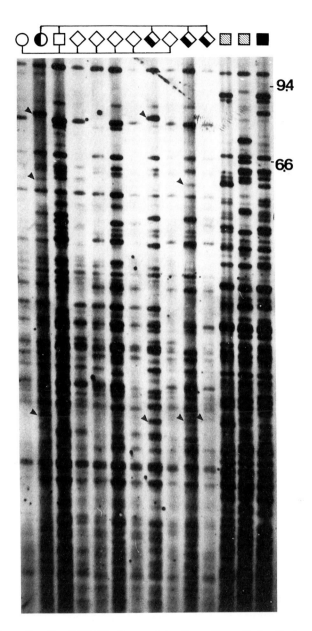

FIGURE 1. Nonradioactive (GGAT)$_4$ fingerprints of a nestbox of zebrafinches with two mothers and one father. The multilocus patterns of the two sisters and their common mate are depicted on the left, those of their jointly reared offspring in the center, and those of three additional, uninvolved males on the right-hand side. The solid black symbol designates the nonpaired male in the same aviary. (For methodological details of DNA preparation, digestion and transfer, and hybridization as well as signal development see Epplen and Mathé 1992, Lubjuhn et al. 1992.) After the initial X-ray film exposure shown here and probe removal, the PVDF-membrane-bound DNA was rehybridized with the radioactively labeled (CA)$_8$ probe (not shown) in order to verify offspring assignation to one of the potential mothers. The arrows mark signal fragments that are diagnostic for maternity.

probes can easily be eluted from the gel by high temperature or a denaturation/renaturation step.

Recent technical developments involving chemoluminescent signal detection on nylon (Amersham or Boehringer Mannheim) or PVDF membranes (Millipore) using digoxigenated oligonucleotides (Fresenius), alkaline phosphatase, and AMPPD (Boehringer Mannheim) have reduced the necessary film exposure times to minutes (see Fig. 1 and also Epplen and Mathé 1992; Lubjuhn et al. 1992). Thus the whole procedure has been accelerated and the use of radioactivity avoided. The fingerprint results are documented on X-ray film, which represents a considerable advantage over the (Polaroid) photos of stained gels or filters after enzymatic signal color development (Zischler et al. 1989). Given the recent establishment of DNA preparation methods from blood requiring less than an hour (Genomix), it is clear that this technique of oligonucleotide fingerprinting has reached a high level of technical sophistication. But the main advantage of the technique is that it is simply foolproof. One major disadvantage of chemoluminescent approaches, however, is that the fingerprints have to be shown on solid supports since the signals are trapped in the agarose gels.

2.6 PCR for Human Autosomal and Y-chromosomal Loci

In order to amplify autosomal and Y-chromosomal loci, 500–1,000 ng of human DNA were used for each PCR reaction. Repeat flanking oligonucleotide primers for the following three loci were synthesized:

Locus 27H39:
27H39.1 5'-CTACTGAGTTTCTGTTATAGT-3',
27H39.2 5'-ATGGCATGTAGTGAGGACA-3'

Locus 4804A:
4804.1 5'-TCATTTAAGCATTTGAGGGAA-3',
4804.2 5'-AGACTTCAAAACAGACACTT-3'

──►

FIGURE 2. Variability of three simple repeat loci as revealed by denaturing polyacrylamide electrophoresis after PCR amplification. Locus 27H39LR originates from the human Y chromosome (one allele, visible in males only), whereas 4804LR and 4815LR stem from human chromosome 12 (two alleles in all persons that are not homozygous). In each part of the figure the length of an allele is indicated in base pairs on the right hand side; in (C) two sequencing reactions have been loaded onto the gel as well. For further methodological details see the Materials and Methods section and Roewer et al. (1992). Note the slippage by four bases in (B), by two to four bases in (A), and by one nucleotide in (C). (A) 1–16 = unrelated male individuals; F = father, S = son of family c or e. (B) F = father, M = mother, and C = children of families b, c, or d. 1, 2, 3, and 4 additional unrelated individuals. (C) 1–15 = independent individuals; F = father, M = mother, C = child of family a.

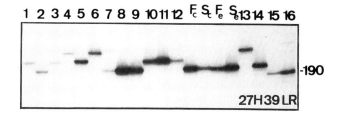

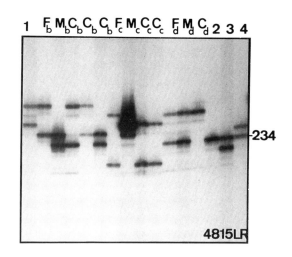

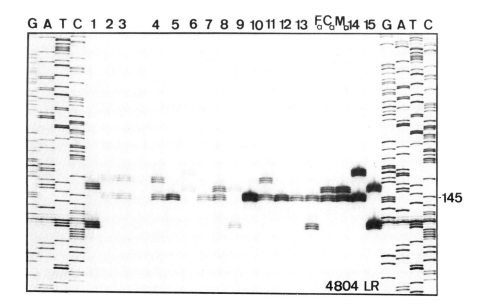

Locus 4815:
4815.1 5′-GCAACAGTTTATGCTAAAGC-3′,
4815.2 5′-GCCTATGCAGTTCAAATCTA-3′

The PCR reactions were carried out according to the conditions recommended by the manufacturer of *Taq* DNA polymerase (Perkin Elmer, Norwalk, CT, USA) at the following temperatures: denaturation at 94°C for 30 sec, annealing at 51°C for 30 sec, and extension at 72°C for 90 sec. Thirty cycles were carried out in a reaction volume of 25 μl (2 mM Mg^{2+}). Simultaneous multilocus PCRs using the three primer pairs were carried out under identical conditions. Gel purification of the repeat-containing fragments, radioactive labeling of the fragments, separation in 4% denaturing polyacrylamide gels, and X-ray film exposure were performed as previously described (Roewer et al. 1991). Mendelian inheritance of the autosomal DNA loci was demonstrated (see also family data in Fig. 2B, C), as well as regular inheritance from fathers to sons with respect to the Y-chromosomal DNA. For exact sizing of the fragment bands, the sequenced clones 27H39, 4804A, and 4815 were used.

3. Applications in Forensic, Animal, Fungal, and Plant Sciences

Multilocus fingerprinting using minisatellite or pure simple repeat probes has led to the first, and numerous subsequent, cases of unequivocal personal identification and determination of genetic relationships (Jeffreys et al. 1985; Nürnberg et al. 1989; for a review see Epplen et al. 1991), not only among human beings but also among individuals of all other eukaryotic kingdoms (Epplen et al. 1991; Meyer et al. 1992; Nürnberg and Epplen 1992). One particular advantage of the oligonucleotide approach consists in the direct in-gel hybridization, allowing the researcher to omit DNA transfer to solid supports. Impressive series of data have been accumulated that prove the practicability, effectiveness, and safety of multilocus systems for routine diagnostic purposes with fresh samples (Jeffreys et al. 1991, Krawczak et al. in press, a). Critical opinions on the procedures for statistical evaluation of the data have been persuasively counterargued (Krawczak et al. submitted, a, b). The fundamental value and the enormous wealth of information produced by this method appear indisputable.

Nevertheless, the attention of quite a few laboratory investigators involved in personal identification (namely those of private enterprises and the FBI) has recently turned to the employment of a collection of single-locus probes. The supposed advantages include the obvious simplicity of the two-signal patterns, their higher sensitivity, and clearer statistics. None of these points holds up too well with respect to most routine applications (e.g., paternity or genetic relationship determinations). In addition, the statistical evaluation procedures for personal identification using monolocus probes have come under serious (scientific and public) pressure recently (Lewontin and Hartl 1991; but see also Chakraborty and Kidd 1991). In forensic science it is the quality of the stain sample that determines whether multilocus systems are applicable and sufficiently informative. Complete degradation of the DNA leaves no adequately long starting molecules for restriction digests (Roewer et al. 1990). This problem, however, arises for both monolocus and multilocus hybridization systems.

In other areas, for example, behavioral research (using previously frozen animal corpses) quite a large number of multilocus fingerprint applications have been profitably pursued, for instance the analysis of mating success in avian (Birkhead et al. 1990) and fish species (Schartl et al. 1991; for studies on ungulates, see e.g. Schwaiger et al. 1992). Extra-pair fertilizations and brood parasitism in birds can be identified with certainty even when the presumed parents are siblings. As just one consequence of this, the "good genes" hypothesis (Gladstone 1979) can now be tested in various immediately accessible bird model systems, and thus be either proven or rejected.

As to clinical applications, multilocus DNA fingerprints have been generated by oligonucleotide probes specific for simple repeats from surgically removed tissue and/or cultured cells of intracranial tumors (gliomas, medulloblastoma, metastatic carcinomas) as well as kidney carcinomas (Epplen et al. 1992). These were compared with the constitutional banding patterns obtained from the peripheral blood leukocytes of each patient. Prominent changes were observed in the $(GACA)_4$ fingerprints of most kidney tumors investigated. It appears that a whole set of rRNA spacers that contain simple $(gaca)_n$ repeat stretches is lost. A multitude of detected somatic changes were found to reflect the chromosome alterations identified by parallel karyotype analysis in the gliomas. Gain and/or loss of bands or significant band intensity shifts could be demonstrated in the fingerprints of more than 80% of the intracranial tumors investigated. These changed patterns were due to highly amplified DNA fingerprint fragments in independent gliomas. Amplification of the epidermal growth factor receptor (EGFR) gene via Southern blot hybridization was only revealed in those tumors showing amplified DNA fingerprint fragments as well. Finally, in transfusion research, deleterious graft-versus-host reactions could be diagnosed "and clarified" by demonstrating the complete absence of white blood cells of the host (Kunstmann et al. 1992).

4. Structural Analysis Based on Hybridization In Situ

The nuclear and chromosomal location of larger amounts of simple repetitive DNA can be visualized by direct hybridization in situ onto nuclear or chromosomal nucleic acids (Zischler et al. 1989). Nonradioactively labeled, digoxigenated probes are preferable to biotinylated or radioactively labeled oligonucleotides (Nanda et al. 1991). The subject of in situ hybridization, for example to metaphase chromosomes, using nonradioactively labeled oligo-nucleotide probes, has been covered in detail in a recent review (Nanda et al. 1991). Suffice it to say here that the limited number of reporter molecules that can be efficiently attached to an oligonucleotide probe certainly limits the sensitivity of the system, so that the analysis of single-copy sequences is not possible for now. More, or more spacious, reporter molecules would interfere with the ability of the probe to hybridize specifically to its target. However, above a threshold amount the accumulation of simple repetitive DNA in a given location can be elegantly visualized with comparatively little effort. The most pertinent advantages of the procedure, besides avoiding radioactivity, include: (a) sharp signals attributable to defined structures; (b) the possibility of multiple different reporter groups for the simultaneous detection of different targets and their spatial relationships; (c) less time-consuming steps (hybridization, exposure, signal development, especially when chemoluminescent reagents such as AMPPD are employed).

The potential of oligonucleotide hybridization in situ has not yet been fully explored. Particular applications concern the accumulation of simple repetitive sequences in defined (chromosomal) locations, for example on the vertebrate gonosomes or on primate homologues of the human acrocentric chromosomes 13–15 and 21–22. In these locations, $(gaca)_n$ elements are interspersed with genes coding for rRNA. The chromosomes in question contain the so-called nucleolus organizer regions (NORs), which can be visualized by silverstaining. Signals vary from individual to individual. Theoretically, there may be an advantage to oligonucleotide hybridizations in the field of ancient DNA research: since, due to degradation, there are only short target molecules left, short probes may be the remedy. But as of today, the direct applications of hybridization in situ in this field appear quite restricted.

5. Single-Copy Probes for Individual Hypervariable Loci from a Human Multilocus Fingerprint

Like Alec Jeffrey's group working with their minisatellite system (Wong et al. 1987), we made every effort to find suitable ways of isolating locus-specific probes from the $(CAC)_5/(GTG)_5$ multilocus fingerprint (Zischler et al. 1991). All "shortcuts" (PCR approaches) proved to be less efficient and not advisable. Finally, four particularly informative single-locus probes were established based on human genomic phage library screening (Zischler et al. 1992). These were purified, characterized, partly sequenced, and finally localized on chromosomes 8, 9, 11, and 22 by way of mouse/human somatic hybrid cells. The results of molecular cloning of those sequences are by no means identical replicas of the genomic situation. But the isolated single-copy parts can be used as probes, though sometimes defined hybridization competitors have to be added to avoid background signals. Using already the derived locus-specific oligonucleotides as probes in the initial investigations, which involved more than 25 defined Eurasian populations, equally broad distributions of allele sizes were demonstrated (Hundrieser et al. 1992): allele heterozygosity rates were well above 90%, even though very conservative assumptions were applied when one single band appeared per individuum. The mutation rate in humans was estimated to be $\frac{0}{200}$ to $\frac{1}{200}$ per locus and meiosis. Interestingly, a pair of monozygotic twins was detected in which one of the brothers exhibited a somatic mosaic for one of the locus-specific probes (Hundrieser et al. 1992).

The production and establishment of experimental conditions for corresponding nonradioactive cloned single-locus probes is currently under way (Gomolka, Schwaiger et al., in prep.).

6. PCR for Simple Repeat Loci

Although even in the mind of the interested public gene technologies show unsurpassed sensitivity, the problems caused by minute amounts of contaminants in PCR amplifications are generally underestimated in the laboratory. Especially in the field of aDNA research, the question of appropriate contamination controls has not been fully solved. Security measures and standards are the subject of various contributions to this book. A mandatory ban on molecular cloning procedures in PCR laboratories appears essential. In addition, it may turn out to be necessary to constantly develop new, highly informative loci to circumvent all contamination problems.

We isolated informative loci carrying tetranucleotide (gata)$_n$ repeats from cosmids derived from human autosome 12 as well as from the Y chromosome (Roewer et al., 1992). The simple-sequence variability was shown to be similar for Y-chromosomal and autosomal loci (see Fig. 2). Since on the (non-pseudo-autosomal part of the) Y chromosome unequal crossing-over events are impossible, this equal variability can be taken as a further argument in favor of *intra*strand mechanisms like slippage being responsible for the generation of different alleles. In the meantime, these loci have already been applied to actual case work involving Caucasians (Roewer and Epplen 1992) and to extended population studies in Brazil (dos Santos et al., 1993). The conditions for multiplex PCR for the simultaneous amplification of three loci have been established. Thus the information accumulated can theoretically be obtained from very limited sample sizes. So far we hesitate to advocate the wide adoption of such PCR systems by laboratories with little DNA amplification experience. Whenever the state of degradation and sample sizes allow a "classical" multilocus fingerprint, that should be procedure of choice. Rules for multilocus fingerprinting of forensic samples using oligonucleotide probes have been formulated and revisited (Roewer et al. 1991).

1 Partial cDNA Sequence Encoding Human *MHC* Class I Promotor Binding Protein 2 Mouse cDNA

```
     601       611       621       631       641       651       661       671       681       691
1    · · · · · · · · · · · · · · · · · · · · · · ·GCAGAGCTTGCTGTGGAACAGAAGAGTGACCAGGGCGTTGAGG
2    GATGGAGATGGGGCTGGGGGAGCCCCTGAGGAGATGCCTGTGGACAGGATCCTGGAG---------G-----------------------A--------

     701       711       721       731       741       751       761       771       781       791
1    GTCCTGGGGAACCGGGGGGTAGCGGCACGAGCCCAAATGACCCTGTGACTAACATCTGTCAGGCAGCTGACAAACAGCTATTCACGCTTGTTGAGTGGGC
2    --------CC---------G-T-----GC----------------------A-----------------------G------A--C----------

     801       811       821       831       841       851       861       871       881       891
1    GAAGAGGATCCCACACTTTCCTCCTTGCCTCTGGATGATCAGGTCATATTGCTGCGGGCAGGCTGGAATGAACTCCTCCTTTTCACACCGA
2    A---------G---C----------C-A--------C-------------------------C--G-------G----C--C--T--G

     901       911       921       931       941       951       961       971       981       991
1    TCCATTGATGTTCGAGATGGCATCCTCCTTGCCACAGGTCTTCACGTGCACCGCAACTCAGCCCATTCAGCAGGAGTAGGAGCCATCTTTGATCGGTCCC
2    ----------C-------------------G------T------A-A-------C------C--G

     1001      1011      1021      1031      1041      1051      1061      1071      1081      1091
1    TCTCCAGGGTGCTGACAGAGCTAGTGTCCAAAATGCGTGACATGAGGATGACAAGACAGAGCTTGGCTGCCTGAGGGCAATCATTCTGTTTAATCCAGA
2    ---------------------------G-----------------------------------C-----------A------

     1101      1111      1121      1131      1141      1151      1161      1171      1181      1191
1    TGCCAAGGGCCTCTCCAACCCTAGTGAGGTGGAGGTCCTGCGGGGAGAAAGTGTATGCATCACTGGAGACCTACTGCAAACAGAAGTACCCTGAGCAGCAG
2    C----------------G-A------A----T-----a-------G------C-C-------------------T----G

     1201      1211      1221      1231      1241      1251      1261      1271      1281      1291
1    GGACGGTTTGCCAAGCTGCTGCTACGTCTTCCTCCGCCCTCCGGTCCATTGGCTTCTAGAGCATCTGTTTTTCTTCAAGCTCATTGGTGACACCC
2    C----------------T--------------------------C---------C-------C----------------C

     1301      1311      1321      1331      1341      1351      1361      1371      1381      1391
1    CCATCGACACCTTCCTCATGGAGATGCTTGAGGCTCCTCCCCATCAACTGGCCTGAGctcagacccagacgtggtgcttctcacactggaggagcacacatcc
2    ----t---------c-a-g---------c---------tta----ac-----a-------g--tt

     1401      1411      1421      1431      1441      1451      1461      1471      1481      1491
1    aagagggactccaagccctgggcgagggtgggggcatgttccccagaaccttgatgggtgagaagtacaggcagaaccaagaacataaaccctcc
2    --gc-tgg--agg---aga-ccat---a---g--g-ag---c-g-------a--g---ctt-cc---cctgcc-ggg-tctggc--ca-t-ag

     1501      1511      1521      1531      1541      1551      1561      1571      1581      1591
1    aagggatctgcttgatatcccaagttggaagggacccagataccctgtgaggactggttgtctcttcggtggccttgagtctcagaatttgtcggggt
2    c---t-- --g-c- -------c-a-g ------------c ---tta----a ----t--

     1601      1611      1621      1631      1641      1651      1661      1671      1681      1691
1    ctccatgatttggggtgatttctcaccctctgtccttacccagcacaaagcactgcctgcctccaggacccttgctccttcctctcatcttgcctcatttt
2    ----catgg-gca-------ct---t----ggc------------c-------------t-g------

     1701      1711      1721      1731      1741      1751      1761      1771      1781      1791
```

```
1 gcttccatctgaagagtggaaatggaactccccagaggtggatactggggcaggcctcccaagctgatggacatgagagtagcgcctgacaggcct
2 --c-- t---------------ca-a-ct-c---ag-a---g-tg-tg-  ----------------c--------at---g---c--g--tctgaca----
  1801    1811    1821    1831    1841    1851    1861    1871    1881    1891

1 tcctcctctcaaacctggcagatggggggctctctggaagagggaggggccctgtcactgtccagagtctcttttttacacttcacctccttctgcagtca
2 -ta----g--t--a----------  -----------a----c-t----tcct---agccccattt-ca--gtc---c------
  1901    1911    1921    1931    1941    1951    1961    1971    1981    1991

1 gactgaaatataaaaaggtggtggtggtgaagggggctggtggagatgtaggaaccgatctgctatttttaatttcctgtgaggatagagacttgca
2 -------ga------g---a-----c----g--g----------c----------------g-g-
  2001    2011    2021    2031    2041    2051    2061    2071    2081    2091

1 gttagactcaaagaagtactgtactttcccagttgactaagaaatgccagtggtggaggtgggtgttgggaaaggcaggccctgaaatggcctgtccc
2 ------------------a-c-----------t ------t---g-c-----a--ca -g---c---ca -g-c-----ag-c----c-gg--g------
  2101    2111    2121    2131    2141    2151    2161    2171    2181    2191

1 tagggctctccaagcactagccttcccagcttcccgcgcccccatctcttcctgtctaacttgggaagggcctgggctgtgaggacaggg ccc
2 a                                                ---ctc----------c---cctcgg-t-g---
  2201    2211    2221    2231    2241    2251    2261    2271    2281    2291

1 cacaggggatggttcacgagtgtagtcccgggaggccttcccttacagctctcctccagccctgggcacatagcatagctgggacacaggatcctgg
2 t----   -t---cc-tag---taag---cag--a---tggg---   --tgg--ac-ttg-t-gggg---
  2301    2311    2321    2331    2341    2351    2361    2371    2381    2391

1 cctgagaattgagggggaggtggccagcccgcagagtgggtgctgtgggctgcatgattttgccctgcgtcccttctctttgggctccttttcccctct
2 --a-cc  -gc--a-tt--gt--tg-g-t---tg-t-ttt-cc---ac--ctt-tggggg-t----c-cat-t--a---gcacataaag--g-tt---c
  2401

1 catacataaaa....
2 ag--ta-a----...
```

FIGURE 3. Comparison of the partial cDNA sequences of the mouse *MHC* class I promotor binding protein (H-2RIIBP, 2) and its human homologous counterpart (1; Epplen and Epplen, 1992). Amplified and unamplified cDNA libraries from total T cell RNAs have been screened for the presence of simple sequence transcripts, using among others the (CAC)$_5$/(GTG)$_5$ probe. A large number of other hybridizing phage clones were purified and partly characterized after subcloning by double-strand plasmid sequence analysis according to the USB Sequenase-2.0 kit. Capital letters represent translated nucleotides.

Dashes indicate identical bases in both cDNAs. Note the presence of *four additional codons* (double underline, bold) around # 1000 in the human sequence as well as that of a perfect (gtg)$_n$ repeat (double underlined) in the human 3'-untranslated sequence only. Putative ligand binding domains (single underline, italic) are moderately conserved. The additional simple (gtg)$_n$ repeat is located immediately after "interspersed" potential poly(A)$^+$ addition signals (single underline, roman) in both sequences.

7. Direct Comparison of Extensive Protein Polymorphism Versus DNA Hypervariability

What can we learn from studies of specific gene loci and the adjacent simple repeats in introns and spacers? At the outset, we wish to reiterate that in many cases the hypervariable simple sequences represent exceedingly useful tools for indirect gene analysis. In fact, they are orders of magnitude more informative than conventional restriction fragment length polymorphism (RFLP) markers. In addition, PCR amplification speeds up the procedure. Thus conventional RFLPs may soon be replaced completely by the analysis of truly hypervariable loci. As a consequence, the complete physical map of, for example, the human genome will soon be available, rendering unnecessary previous heroic efforts that required innumerable hands and time (and money).

In most instances, simple repeats arise and vanish as exemplified in the transspecies comparison of the 3'-untranslated regions of the Major Histocompatibility Complex (MHC) class I promotor binding protein genes (Epplen and Epplen, 1992, see Fig. 3; the mouse sequence has been taken from Hamada et al. 1989). The mouse cDNA also shows clear-cut homology in parts of the 3'-untranslated regions. Yet for unknown reasons the simple $(gtg)_n/(cac)_n$ repeat blumes only in the human transcribed sequence (Epplen and Epplen, 1992).

Proving most informative up to now with respect to the comparison of protein polymorphism versus DNA hypervariability has been the MHC class II DRB locus, which encodes one chain of a transplantation antigen that is involved in antigen presentation to T lymphocytes: adjacent to the highly polymorphic exon 2 sequence, intronic simple $(gt)_n/(ga)_m$ repeats allow for the classification of the respective exon on the basis of its basic architecture. This is not only the case in humans (Rieß et al. 1990), but also found for the very same location in farm animals like cattle or goats, and in other even-toed ungulates including gazelles and giraffes (Ammer et al. 1992; Schwaiger et al., in prep.). Detailed studies unexpectedly revealed that exonic and intronic sequences evolve at approximately the same rate—in stark contrast to the situation described above for the MHC class I promotor binding protein. Does the mixed simple repeat stretch induce particular secondary structures? Does it represent a special landmark within the chromatin? We believe there is a lot more to be learned from parallel studies on ancient samples.

Our protein binding studies of these mixed simple repetitive $(gt)_n(ga)_m$ sequences have revealed that there are nuclear factors targeted specifically toward these elements or their secondary structure (Mäueler et al. 1992). Interestingly, we have identified a factor clearly different from that of Yee et al. (1991) whose object of study binds exclusively to single-stranded nucleic acids. We expect to see rapid development of this unfolding area of interdisciplinary research.

8. Prospect

With respect to the biological origin and role of repetitive DNA sequences, it appears that several "DNA philosophers" have developed plausible hypotheses (for discussion, see e.g. Dover 1991). Having became disillusioned in recent years with attempts to unravel the secrets of these elements in due time, we feel that researchers in this field limit themselves to conventional concepts. Nevertheless, the systematic and unbiased analysis of particularly informative simple repeat loci that have been preserved over considerable evolutionary time spans may reveal anything one might want to know about these elements. In addition, the handiness of simple repeats as tools has opened unexpected avenues for a variety of biological diagnostic disciplines. It would be surprising if studies on aDNA were an exception. The consensus is that PCR-amplification-based systems are the only appropriate means to investigate (highly) degraded nucleic acid remains from old biological specimens. Thus, if PCR "hypersensitivity" can be sufficiently controlled for, hypervariable simple repeat loci might well do the job in all those studies of aDNA involving questions of individualization, whether in plants, fungi, or animals including humans.

Acknowledgments. The results described herein have been obtained in collaboration with J. Buitkamp, C. Epplen, M. Gomolka, T. Lubjuhn, W. Mäueler, L. Roewer and F. Schwaiger. The expert secretarial assistance of Heidi Sommerfeld is highly appreciated, as is the technical help by Irene Bergmann, Simone Quentemeier, and Renate Steppke. Work described in this chapter has been supported by grants from the VW-Stiftung, the Sander-Stiftung and the DFG (Ep 7/5-2; Ep 7/6-2). The fingerprinting probes are subject to patent applications. Commercial inquiries should be directed to Fresenius AG, W-6370 Oberursel, Germany.

References

Ammer H, Schwaiger FW, Kammerbauer C, Gomolka M, Arriens A, Lazary S, Epplen JT (1992) Exonic polymorphism vs intronic simple repeat hypervariability in *MHC-DRB* genes. Immunogenetics 35:330–337

Birkhead TR, Burke T, Zann R, Hunter FM, Krupa AP (1990) Extra-pair paternity and intraspecific brood parasitism in wild zebra finches *Taeniopygia guttata*, revealed by DNA fingerprinting. Behav Ecol Sociobiol 27:315–324

Chakraborty R, Kidd KK (1991) The utility of DNA typing in forensic work. Science 254:1735–1739

Dover G (1991) From philosophy to forensics. Fingerprint News 3(4):6–7

Epplen JT (1988) On simple repeated GATA/GACA sequences: a critical reappraisal. Heredity 79:409–417

Epplen JT (1992) The methodology of multilocus DNA fingerprinting using radioactive or nonradioactive oligonuleotide probes specific for simple repeat motifs. In: Chrambach A, Dunn MJ, Radola BJ (eds) *Advances in Electrophoresis*, Vol. 5. Weinheim: Verlag Chemie, pp. 59–112

Epplen C, Epplen JT (1992) The human cDNA sequence homologous to the mouse *MHC* class I promotor binding protein gene contains four additional codons. Mammalian Genome 3:472–475

Epplen JT, Mathé J (1992) Multilocus DNA fingerprinting using nonradioactively labeled oligonucleotide probes specific for simple repeat elements. In: Kessler C (ed): *Non-radioactive labelling*, Heidelberg: Springer-Verlag, 271–277

Epplen JT, Ammer H, Epplen C, Kammerbauer C, Mitreiter R, Roewer L, Schwaiger W, Steimle V, Zischler H, Albert E, Andreas A, Beyermann B, Meyer W, Buitkamp J, Nanda I, Schmid M, Nürnberg P, Pena SDJ, Pöche H, Sprecher W, Schartl M, Weising K, Yassouridis A (1991) Oligonucleotide fingerprinting using simple repetitive repeat motifs: a convenient, ubiquitously applicable method to detect hypervariability for multiple purposes. In: Burke T, Dolf G, Jeffreys AJ, Wolff R (eds) *DNA-Fingerprinting: Approaches and Applications*. Basel: Birkhäuser-Verlag, pp. 50–69

Epplen JT, Bock S, Nürnberg P (1993) Tumor genome screening by multilocus DNA fingerprints as obtained by simple repetitive oligonucleotide probes. In: Wagener C, Neumann R (eds) *Molecular Diagnostics of Cancer* 41–51

Gladstone DE (1979) Promiscuity in monogamous colonial birds. The American Naturalist 114:545–557

Hamada K, Gleason SL, Levi B-Z, Hirschfeld S, Appella E, Ozato K (1989) H-2RIIBP, a member of the nuclear hormone receptor superfamily that binds to both the regulatory element of major histocompatibility class I genes and the estrogen response element. Proc Nat Acad Sci USA 86:8289–8293

Hundrieser J, Nürnberg P, Czeizel A, Metneki J, Rothgänger S, Foelske C, Zischler H, Epplen JT (1992) Characterization of hypervariable, locus specific probes derived from a (CAC)$_5$/(GTG)$_5$ fingerprint in various Eurasian populations. Hum Genet 90:27–33

Jeffreys AJ, Wilson V, Thein SL (1985) Hypervariable 'minisatellite' regions in human DNA. Nature 314:67–73

Jeffreys AJ, Wilson V, Neumann R, Keyte J (1988) Amplification of human minisatellites by the polymerase chain reaction: towards DNA fingerprinting of single cells. Nucl Acids Res 16:10953–10971

Jeffreys AJ, Turner M, Debenham P (1991) The efficiency of multilocus DNA fingerprinting probes for individualization and establishment of family relationships determined from extensive casework. Am J Hum Genet 48:824–840

Krawczak M, Böhm I, Nürnberg P, Hundrieser A, Pöche H, Peters C, Slomski R, Pöpperl A, Epplen JT, Schmidtke J (in press) Paternity testing with oligonucleotide probe (CAC)$_5$/(GTG)$_5$: a multi-center study. Forens Sci Int

Kunstmann E, Bocker T, Sauer H, Mempel W, Epplen JT (1992) Diagnosis of transfusion-associated graft-versus-host disease by genetic fingerprinting and polymerase chain reaction. Transfusion 32:776–770

Lewontin RC, Hartl DL (1991) Population genetics in forensic DNA typing. Science 254:1745–1750

Lubjuhn T, Curio E, Epplen C, Epplen JT (1992) Non-radioactive oligonucleotide fingerprints using AMPPD reveal a case of unusual reproductive strategy in Timor zebra finches (*Taeniopygia guttata guttata*). Fingerprint News 4(2):13–14

Mäueler W, Muller M, Köhne AC, Epplen JT (1992) A gel retardation assay system for studying protein binding to simple repetitive DNA sequences. Electrophoresis 13:7–10

Meyer W, Lieckfeldt E, Kayser T, Nürnberg P, Epplen JT, Börner T (1992) Fingerprinting fungal genomes with phage M13 DNA and oligonucleotide probes specific for simple repetitive DNA sequences. Adv Mol Genet 241–253

Nanda I, Epplen JT, Schmid M (1991) *In situ* hybridization of nonradioactive oligonucleotide probes to chromosomes. In: Adolph KW (ed) *Advanced Techniques in Chromosome Research.* New York: Marcel Dekker, pp. 117–134

Nürnberg P, Epplen JT (1992) On the reusability of dried agarose gels for the subsequent analysis of multiple samples from various kingdoms. Fingerprint News 4(1):9

Nürnberg P, Roewer L, Neitzel H, Sperling K, Pöpperl A, Hundrieser J, Pöche H, Epplen C, Zischler H, Epplen JT (1989) DNA fingerprinting with the oligonucleotide probe $(CAC)_5/(GTG)_5$: somatic stability and germline mutations. Hum Genet 84:75–78

Rieß O, Kammerbauer C, Roewer L, Steimle V, Andreas A, Albert E, Nagai T, Epplen JT (1990) Hypervariability of intronic simple $(gt)_n(ga)_m$ repeats in *HLA-DRB* genes. Immunogenetics 32:110–116

Roewer L, Epplen JT (1992) Rapid and sensitive typing of forensic stains by PCR amplification of polymorphic simple $(gata)_n$ repeats. Forensic Sci Int 53:163–171

Roewer L, Nürnberg P, Fuhrmann E, Rose M, Prokop O, Epplen JT (1990) Stain analysis using oligonucleotide probes specific for simple repetitive DNA sequences. Forensic Sci Int 47:59–70

Roewer L, Rieß O, Prokop O (1991) Hybridization and polymerase chain reaction amplification of simple repeated DNA sequences for the analysis of forensic stains. Electrophoresis 12:181–186

Roewer L, Arnemann J, Spurr NK, Grzeschik KH, Epplen JT (1992) Simple repeat sequences on the human Y chromosome are equally polymorphic as their autosomal counterparts. Hum Genet 88, 89:389–394

dos Santos FR, Pena SDJ, Epplen JT (1993) Genetic and population study of a y-linked tetranucleotide repeat DNA polymorphism with a simple non-isotopic technique. Hum Genet 90:655–656

Schäfer R, Zischler H, Birsner U, Becker A, Epplen JT (1988) Optimized oligonucleotide probes for DNA fingerprinting. Electrophoresis 9:369–374

Schartl M, Erbelding-Denk C, Nanda I, Schmid M, Schröder JH, Epplen JT (1991) Mating success of subordinate males in a poeciliid fish species, *Limia perugiae.* Fingerprint News 3(2):16–18

Schwaiger FW, Gomolka M, Geldermann H, Zischler H, Buitkamp J, Epplen JT, Ammer H (1992) Oligonucleotide fingerprinting to individualize ungulates. Theor Appl Electrophoresis 2:193–200

Singer MF (1982) SINES and LINES: highly repeated short and long interspersed sequences in mammalian genomes. Cell 28:433–434

Skinner D (1977) Satellite DNAs. Bioscience 27:790–796

Tautz D, Renz M (1984) Simple sequences are ubiquitous repetitive components of eukaryote genomes. Nucl Acids Res 12:4127–4137

Weber JL (1990) Human DNA polymorphisms based on length variations in simple-sequence tandem repeats. In Davis KE, Tilghman SM (eds), *Genome Analysis 1: Genetic and Physical Mapping.* Cold Spring Harbor, N.Y.: CSH Laboratory Press, pp. 159–181

Wong Z, Wilson V, Patel I, Povey S, Jeffreys AJ (1987) Characterization of a panel of highly variable minisatellites cloned from human DNA. Ann Hum Genet 51:269–288

Yee HA, Wong AKC, van de Sande JH, Rattner JB (1991) Identification of novel single-stranded d(tc)n binding proteins in several mammalian species. Nucl Acids R 19:949–953

Zischler H, Nanda I, Schäfer R, Schmid M, Epplen JT (1989) Digoxigenated oligo-nucleotide probes specific for simple repeats in DNA fingerprinting and hybridization in situ. Hum Genet 82:227–233

Zischler H, Hinkkanen A, Studer R (1991) Oligonucleotide fingerprinting with (CAC)$_5$: Nonradioactive in-gel hybridization and isolation of individual hypervariable loci. Electrophoresis 12:141–146

Zischler H, Kammerbauer C, Studer R, Grzeschik K-H, Epplen JT (1992) Dissecting (CAC)$_5$/(GTG)$_5$ fingerprints from man into individual locus specific, hypervariable components. Genomics 13:983–990

Evolutionary Analysis
3 Spatial and Temporal Aspects of Populations Revealed by Mitochondrial DNA

Francis X. Villablanca

1. Introduction

The evolutionary analysis of DNA sequences bridges phylogenetics and population genetics. Ancient DNA (aDNA) allows the study of extinct genotypes, populations, and species, as well as dichronic comparisons of extant populations and species. Thus aDNA forges an empirical link between history and two inherently historical fields of research. Fortunately, the conceptual frameworks of phylogenetics and population genetics can easily be extended to encompass advances being made in the study of aDNA.

In order to demonstrate and discuss the evolutionary analysis of ancient DNA sequences, we will be investigating an expanded version of the study originally presented by Thomas et al. (1990). Thomas and coworkers sequenced 225 base pairs of the mitochondrial control region (d-loop) from 43 museum specimens and 63 modern specimens of the kangaroo rat *Dipodomys panamintinus* (Rodentia, Heteromyidae). In total, the 106 specimens represent two temporal samples from each of three different geographical populations. Modern samples were taken in 1988, 53 to 77 years after the museum specimens were collected. Expanding this study, we will consider a total of 421 base pairs from 14 modern and 5 museum specimens. These 19 specimens represent unique mitochondrial genotypes found by Thomas et al. and will be used to better resolve the evolutionary relationships among genotypes and quantify gene flow between populations.

To date, *D. panamintinus* is the only species studied for which population size samples of aDNA have been applied toward phylogenetics and population genetics. From this perspective it is exemplary. Yet it should be noted that the results are still tentative and await sequencing of longer fragments and a better understanding of intraspecific geographic variation in *D. panamintinus*. In the sections below we consider methods for reconstructing hypothesized evolutionary relationships among genotypes, populations, and species. The hypotheses include estimates of gene flow between populations and of time since divergence between populations. We also quantify genetic variability at the nucleotide level, and measure its change over time. The discussion of each of these approaches will be followed immediately by an application to the data from kangaroo rats. Before addressing phylogenetics and population genetics, let us consider the genome under investigation.

2. Mitochondrial DNA and Gene Trees

The mitochondrial genome has been used extensively in evolutionary studies of animals. Probably the single most important reason for its widespread use is its rate and mode of evolution (Aquadro et al. 1984; Wilson et al. 1985; Moritz et al. 1987; Miyamoto and Boyle 1989). Population level studies exploit the genome's rapid evolutionary rate (Brown et al. 1979, 1982); substitutions accumulate rapidly in third codon positions of coding genes and in the control region (Kocher and Wilson 1991). Interspecific and higher level relationships can be estimated using the slower amino acid replacement changes (Brown 1985; Moritz et al. 1987; DeSalle et al. 1987). Protein coding genes may be particularly useful at bridging the intra- and interspecific levels. Unfortunately, plant mitochondrial DNA evolves faster in structure than in point mutations (Wolfe et al. 1987), and as a result evolutionary studies of plants make use of the chloroplast genome instead (Clegg et al. 1991; Golenberg 1991).

Unlike nuclear DNA, the mitochondrial genome is effectively haploid. Its inheritance is generally considered to be strictly maternal (Wilson et al. 1985; Gyllensten et al. 1991). This must be taken into account as we consider what questions can feasibly be asked when employing the mitochondrial genome (Avise et al. 1987; Birky et al. 1983, 1989; Birky 1991). Specifically, we can only draw conclusions with regard to maternal phylogenies and maternal gene flow. Mitochondrial genome patterns may be quite different from those shown by nuclear genes, particularly if the mating or dispersal/philopatry strategy differs strongly between the sexes. In the absence of sex-specific differences, however, the patterns should be like those from any randomly chosen nuclear gene.

Due to its haploid nature the mitochondrial genome does not undergo recombination (Wilson et al. 1985). This is especially important from the perspective of phylogenetics, which generally assumes that gene evolution involves mutation and branching but no reticulation or recombination of gene fragments. Thus, the mitochondrial genome is considered one completely linked locus. An implication of linkage is that selection acting on one gene affects other genes via hitchhiking (Kaplan et al. 1989; Kreitman 1991).

The mode of inheritance and evolution of mitochondrial DNA will both help and hinder an investigation of aDNA. A practical consideration of great importance is that the mitochondrial genes occur in much higher copy numbers than nuclear genes. As a result, it may be over 100 times easier to amplify a mitochondrial sequence. A high copy number also means that the consensus sequence produced by the polymerase chain reaction (PCR) is derived from a larger number of templates, thus avoiding errors due to template-specific degradation (Pääbo and Wilson 1988). Copy number and retrieval advantages are offset by the fact that mitochondrial DNA is a single locus. Theoretical considerations indicate that several loci need to be studied in order to confidently recover phylogenies (Pamilo and Nei 1988). One locus produces a gene tree. It cannot be expected that gene trees will always be equivalent to species trees, except under specific demographic conditions (Takahata and Slatkin 1990). Thus gene phylogenies from ancient mitochondrial DNA may be easy and reliable, but without phylogenies from nuclear genes such gene trees are only tentative hypotheses of phylogenetic relationships (especially where males and females differ in life history). It is also necessary to distinguish between gene trees and population trees: our analysis will resolve the dichotomous relationships among alleles, but caution is needed in interpreting relationships among populations, which may be reticularly related due to the continued exchange of alleles.

3. Phylogenetics

Let us consider the construction of phylogenetic hypotheses and their applications (reviewed by Swofford and Olsen 1990; Felsenstein 1979, 1988; Wiley 1981). The basic questions are: how are the observed mitochondrial genotypes related genealogically, and how do these relationships compare to the geographic distribution of genotypes. Prior to interpreting our results, we need to evaluate the information content of the nucleotide sequences.

3.1 Data and Samples

D. panamintinus occurs primarily around the western periphery of the North American Mojave Desert. The three kangaroo rat populations studied by Thomas et al. (1990) are referred to as A, B, and C, and represent samples from three discontinuous portions of the species' range. These authors present separate phylogenies for the modern and museum genotypes, demonstrating that the results are essentially equivalent. For simplicity, and because most researchers will be combining the two types of data, we will consider a combination of modern and ancient genotypes. All unique modern genotypes will be included ($n = 14$), as will three types found only in museum samples of population B and two samples of a type which, based on 225 base pairs (Thomas et al. 1990), is identical and shared between museum populations B and C.

Extractions were accomplished in solutions of 5% Chelex (Walsh et al. 1991). PCR from liver tissue was performed in either 25- or 50-μl reactions containing 67 mM Tris (pH 8.8), 16.8 mM AmSO$_4$, 2 mM MgCl$_2$, 01 M β-mercaptoethanol (final concentration), and 0.5 units of *Taq* polymerase. Primers were kangaroo rat specific (TAS; 5'-CTTGCATGTTCATCGTGC-ATT-3') and generalized vertebrate d-loop primers (H16498; Kocher and Wilson 1991) at 1 mM final concentration. Museum skins were amplified using 10 mM Tris (pH 8.3), 50 mM KCl, 1.5 mM MgCl$_2$, 0.001% gelatin, 1.25

units of *Taq*, 2 µg/ml bovine serum albumin, and 40 rather than 30 PCR cycles. A second kangaroo rat specific primer (Dpd4; 5'-TACATTAGAC-ATTAATCCACATAA-3') which amplified a shorter fragment (270 nucleotides rather than 375) was used in combination with H16498 when amplifying from skins. Thermocycling consisted of denaturation at 92°C (60 sec), annealing at 55°C (45 sec), and extension at 72°C (60 sec). All products were sequenced directly in both directions. Figure 1 shows all nucleotide states for variable positions found across 421 base pairs.

Samples of aDNA pose two potentially significant problems to phylogenetic analysis. The first is contamination. Extreme care must be taken to ensure that the relationships being hypothesized are based on amplification of target DNA rather than of contaminants. This can be done simply by limiting the possibility of cross-contamination (Orrego 1990) and by sequence alignment

```
             11111111112222222222222222233334
             14244577899900112333334899916680
             30044720293452503001256015601449
A1.    AAGCGTAGCTGCCGACAATTTAACGGACTTGG
A2.    ACACACCACCACCGATAACTTAACGGACCTGA
A3.    AAGCATGGCTGCCGACAATTTCACGTACTTGG
A4.    AAGCATCACTGCCGATAATTTAATGGACTTGA
A5     AAGCATCGCTGCCGACGATTTAACGGACATGN
A6.    AAGCATAGCTGCCGACAATTTAGCAGTCTCGA
A7.    AAGCATAACTGTCGACAATTTAACGGACCTCA
B1.    AAGTATCACCACCAATAATTTAACGGACCTGA
B2.    GAGTATCACCAACAATAATTTAACGGACCTGA
B3.    AAGTATCACTACCAATAATTTAACGGATCTGA
B4.    AAGTATCACCACCGATAATTTAACGGACATGA
B5.    AAGTACCACCACCAATAATTTAACGGACATGA
B6.    AAGTATCACCACCAGTAATTTAACGGACATGA
B7.    AAGTATCACCACCAATGATTTAACGGACATGA
C1.    AAGCATCATCGCTGATAGCCAAACGGACCTGA
C2.    AAGCATCGTCGCTGATAGCCAAACGGACCTGA
C3.    AAGCATAACTGCCGACAACTTAACAGACTTGG
C4.    ACGCATCATCGCTGATAATTTAACGGACCTGA
C5.    AAGTATCACCACCAGTAATTTAACGGACATGA
D.A.   ???TATAATTA_CAACAACTCAACAGA_????
D.H.   ???TATAACTACCGACAACTTAACAGA_????
```

FIGURE 1. Character states of the 32 variable nucleotide positions from a 421-bp fragment of *Dipodomys panamintinus* control regions. Letters on the left designate the three populations sampled, while letter–number combinations identify unique mitochondrial genotypes within populations. Specimens examined are the first representatives of each genotype as presented in the appendix of Thomas et al. (1990). D.A. and D.H. are the outgroups, *D. agilis* and *D. heermanni*, respectively. Insertions/deletions are indicated by a dash, and undetermined sequences with a question mark. The first variable position is 3, and the last is 409. Position 1 corresponds to position 15307 of the mouse mitochondrial genome (Bibb et al. 1981).

and comparison of amplified DNA with that of known individuals (Thomas et al. 1989). Knowledge of modern levels of variation within a group of organisms will be helpful to assess whether aDNA sequences fall within an expected range of variation. The use of negative controls (extractions and amplifications into which no DNA is added) will reduce the chance that amplification of contaminants will go unnoticed (Pääbo 1990; Thomas et al. 1990). For some primer/template combinations, particularly if primer specificity is poor or if primers are "universal," UV irradiation may also be necessary (Golenberg 1991).

The second potential problem is diagenetic modification resulting in degradation of DNA (Pääbo 1989, 1990). If degradation is randomly scattered across a gene, then the fact that an amplified fragment is a consensus of multiple sequences should compensate for it (Pääbo and Wilson 1988). However, if only one target template is present in the amplification, then any diagenetic modifications will appear in the final sequence. One possible solution (though probably time consuming) is to amplify several highly diluted extractions containing on average zero to one target template each (cf. Golenberg 1991). An alternative is to sequence several extractions per individual. Another alternative (Golenberg 1991) exploits the integrity of double-stranded DNA and repairs the templates prior to amplification. In all of these cases, differences among sequences derived from a single individual could be considered diagenetic modifications. Importantly, diagenetic changes may be less of a problem when considering phylogenetics rather than genetic diversity. Diagenetic changes will likely appear as uniquely derived substitutions (see below); consequently they will have no effect on phylogenetic groupings.

If, on the other hand, the same diagenetic modifications were to occur systematically across samples (i.e., strand breakage and jumping PCR with an adenine insertion at exactly the same site; Pääbo et al. 1990), such modifications would be considered a *synapomorphy* (defined below). The likelihood of this may be small, as is the likelihood that only one character would greatly alter a phylogeny. But, however remote, the possibility does exist.

In the case of *D. panamintinus*, several precautions were taken. Negative controls were used throughout as were cross contamination precautions. At least one kangaroo rat specific primer was always used. PCR conditions were optimized to ensure efficient production of double-stranded product, and therefore a consensus of multiple sequences. Ultimately, sequence alignments, patterns of variation, and phylogenetic placement were used to control for authenticity.

3.2 Phylognetic Analyses and Results

3.2.1   Homology, Saturation, Synapomorphies, and Outgroups

In order for nucleotide sequences to contain information regarding evolutionary relationships, the character states present in any one position must be *homologous*. That is, they must be shared due to ancestry or be identical by descent. We can begin by assuming that homology is the basis of similarity (Hennig 1965), but we will need to evaluate this assumption. We may be able to reject homology if we can demonstrate that molecular evolutionary change is saturated (Villablanca and Thomas 1993; Hillis 1991; Hillis and Huelsenbeck 1992).

Molecular evolution tends to follow a single model when we consider one gene and one group of organisms (Brown et al. 1982; Brown and Simpson 1982; Aquadro et al. 1984). For example, vertebrate mitochondrial DNA evolves with a bias toward transitions relative to transversions (ibid). If observed replacements deviate extensively from the model, we conclude that some actual changes are going unobserved, as a result of evolution in parallel or via reversals (Brown et al. 1979). Mutational differences will appear to plateau once they reach a point of saturation. Importantly, saturated characters can no longer be considered homologous (Villablanca and Thomas 1993). Figure 2 shows that in kangaroo rat control regions, transitions make up the majority of mutations until the pairwise sequence difference between any pair of sequences reaches about 5%. Beyond that point, transitions continue to occur, yet there is no record of their continued evolution since transversions will begin to be counted instead. The genotypes from *D. panamintinus* considered below differ by a maximum of 2.76% corrected sequence divergence (observed sequence difference: max = 2.7%, mean = 2.06%, st. dev. = 0.6). This does not imply that all transitions are homologous, but it does suggest that as a category of change, transitions will be informative.

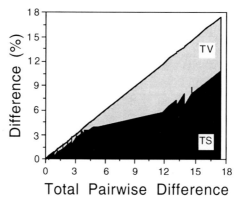

FIGURE 2. Difference in percentage contributed by transitions (TS) and transversions (TV) as a function of total (TS + TV) pairwise difference, indicating the level of pairwise difference at which a transition-biased model is no longer supported. Control region data are from six species of kangaroo rats, including population samples of *D. panamintinus*.

The grouping of organisms based on synapomorphies was one of the most significant conceptual advances of early phylogenetic systematics (Hennig 1965). We can divide evolutionary homologies into shared ancestral, shared derived, and unique (unshared). Only synapomorphies, or derived characters shared between one ancestor and its descendants, will be indicative of phylogenetic relationship. In contrast, shared ancestral characters occur in the ancestors of several mutually exclusive groups and are useless for identifying evolutionary relationships.

Outgroups are used to distinguish ancestral from derived character states (Watrous and Wheeler 1981; Maddison et al. 1984). An outgroup, comprised of at least two species from the sister group, or a more distantly related group, is used to root a phylogenetic tree. Character states shared between the ingroup and the two or more species in the outgroup are considered to be ancestral. Mutations from these character states to some new states would represent shared derived characters of the ingroup.

Phylogenetic studies of relationships among kangaroo rat species (Villablanca and Thomas 1993) using a 432-base-pair fragment of the mitochondrial cytochrome *b* gene (excluding silent first and third position transitions; see also Irwin et al. 1991, Cracraft and Helm-Bychowski 1991), led to the conclusion that the sister group to *D. panamintinus* is a clade containing at least four species. Here we will employ d-loop data from two of these species, *D. agilis* and *D. heermanni*, as an outgroup. Comparison with Thomas et al. (1990) shows that changing the outgroup to one species (*D. californicus*) results in only a minor change among branches that are not fully resolved to begin with. Thus, we are confident that either outgroup will yield similar results; by employing two species in the outgroup we decrease the probability of chance similarity between the outgroup and *D. panamintinus*.

The phylogenetic hypothesis presented here is based on the principle of *parsimony*. Swofford and Olsen (1990) provide a thorough and unbiased review of this and alternative methods for reconstructing phylogenies. Parsimony analysis does not explicitly assume that evolution occurs parsimoniously (all changes at all positions equally likely) but that we should accept the simplest explanation of the data (Wiley 1981, p. 111). As a result, the tree chosen under the parsimony criterion is the shortest tree—the one requiring the fewest number of events (observed and implied) to explain the distribution of characters among the organisms studied (Farris 1983). Fitch, or multistate unordered, parsimony is commonly used for nucleotide analysis (Felsenstein 1988). Though some proponents of parsimony say that its strength lies in the absence of the need to infer a specific model of evolution, we wish to emphasize that the strength of unweighted parsimony is seen when all changes occur at a low and equal rate and when all positions are equally likely to change (Felsenstein 1983, 1988; Villablanca and Thomas 1993). This is the model that parsimony inherently assumes, and the one under which it gives the most reliable results (Felsenstein 1983). In order to incorporate models of molecular evolution into evolutionary analysis, the computer program PAUP (version

3.0; Swofford 1989) allows the user to input a transition/transversion matrix weighting character state changes, as well as to assign weights to different positions. Interest in maximum likelihood methods (PHYLIP; Felsenstein 1983, 1988, 1989) is currently on the rise, primarily due to their greater statistical power, inherent acceptance of explicit models of evolution, and applications to population genetics (ibid: Hudson 1990). The acceptance of weighted parsimony analysis will likely pave the way for increased use of maximum likelihood methods.

3.2.2 Heuristic Parsimony Analyses and Consensus Trees

As the number of individuals being analyzed increases, so does the number of possible trees which need to be evaluated. Theoretically one should look at all possible trees, determine the number of changes required by each, and then choose the shortest tree. For the 22 individuals under consideration here we would have 3.2×10^{23} possible trees (Felsenstein 1978). In order to avoid a computational nightmare, PAUP performs a *heuristic search* (Swofford and Olsen 1990). This is an estimation of the shortest tree, obtained by starting with one individual and adding the others such that the tree length is minimized, then switching branches until one or more shorter or equally parsimonious trees are found, in which case branch swapping begins all over. One replicate of a heuristic search will produce all of the shortest trees which result from branch swapping on a given starting tree. This procedure is replicated ten or more times, with individuals being added to the starting tree in random order for each replicate.

In *D. panamintinus*, 32 of the 421 nucleotide positions vary across genotypes. Twenty of these positions have states shared by two of more genotypes and are thus inferred to be phylogenetically informative. Twelve positions have characters which occur uniquely in only one genotype. Ten replicated heuristic searches resulted in one set of 325 equally parsimonious trees. These trees are summarized in the *consensus tree* presented in Figure 3. The advantage to summarizing data in this way is that if any one of the 325 trees does not resolve a branch, then the branch will appear as part of a polycotomy in the consensus tree (Swofford 1989). The disadvantage is that it suggests some trees which are not observed in the heuristic results, and it conveys less information than any one of the equally parsimonious trees (Carpenter 1988).

Several reasons may exist for lack of phylogenetic resolution. First, the data may not have enough characters to resolve all relationships. This is certainly a problem with *D. panamintinus*, since we are trying to resolve 20 internal branches with exactly 20 "informative" characters. Second, characters may conflict with one another due to parallelisms or reversals (jointly termed *homoplasy*). Homoplastic characters can have several equally parsimonious resolutions. Of the 20 "informative" characters, 7 demonstrate homoplasy. Third, some of the branches connecting genotypes may be very short, with few or no substitutions supporting them. If this problem is suspected, more

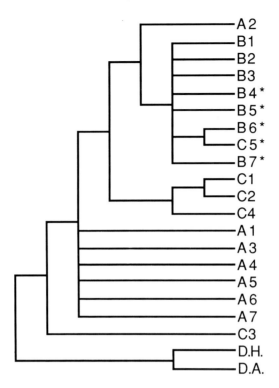

FIGURE 3. Strict consensus of 325 equally short trees obtained from 10 heuristic maximum parsimony searches of the data in Figure 1. All 10 replicates produced the same set of 53-step trees. Asterisks identify genotypes sequenced from museum specimens.

sequence data need to be collected. As indicated at the beginning of this chapter, our results are tentative for exactly this reason.

3.2.3 Exhaustive Parsimony Analyses and Tree Length Distributions

In some cases a disproportionate amount of the homoplasy may be contributed by the outgroup. If the outgroup's level of divergence relative to the ingroup places it in the region of saturation, we may suspect that at least some of the similarities are due to homoplasy rather than homology. Given that in this *D. panamintinus* data the mean sequence difference between the ingroup and the outgroup is 15.6%, we may be better off using a *functional outgroup* (Watrous and Wheeler 1981; Maddison et al. 1984). Regardless of how we root the tree, the relationship between *D. heermanni, D. agilis,* and genotype C3 (Fig. 3) is consistent. This relationship holds up even when the tree is rooted using only transversions between the ingroup and *D. merriami* and *D. nitrotoides,* which differ from the ingroup by a mean sequence difference of 22%. Thus it is safe to conclude that genotype C3 would function as an excellent root (outgroup) for the remainder of the genotypes (ingroup). Making C3 the functional outgroup allows a reduction of the homoplasy resulting from the use of too divergent an outgroup.

An *exhaustive search* differs from a heuristic search in that we calculate the length of all possible trees (Swofford 1989). This procedure can currently be conducted for up to 11 genotypes and guarantees finding the shortest tree(s). Above 11 one can analyze random subsets of genotypes (Hillis 1991). An advantage of exhaustive searches is that we can output and analyze the frequency distribution of tree lengths. Hillis (1991) has shown that the distribution of tree lengths and the skewness of this distribution can be used to determine the relative information content of DNA sequences. A data set of randomized sequences will produce trees with a symmetrical length distribution. Data that contain significant phylogenetic information will result in very few trees toward the optimal (shorter) end of the distribution, and thus a skew toward that tail. The level of skewness can be determined by the g-statistic. The probability of observing the shortest tree can be inferred from the mean and standard deviation of tree lengths. These statistics are provided by PAUP along with the results of an exhaustive search.

The *D. panamintinus* data have been analyzed by three separate exhaustive searches. Each of the three populations is compared to the two others (A/B, A/C, B/C); the resultant trees and the length distributions are presented in Figure 4. In all cases, most parsimonious trees are between 4 and 5.6 standard deviations from the mean, with significant ($p < 0.01$) g-statistics (Hillis and Huelsenbeck 1992). The analysis involves comparison of 10 to 11 genotypes, using 12 to 15 phylogenetically informative positions. Increasing the number of individuals and the sequence length will add statistical power to this test (cf. Hillis and Huelsenbeck 1992).

The phylogenetic results permit several conclusions regarding *D. panamintinus* population structure. But first we must define the terms monophyly, paraphyly, and polyphyly (Wiley 1981 sensu Farris). *Monophyly* means that two groups have no genotypes in common; the groups arise from two separate ancestors. If we were to ignore genotype C5, Figure 4.3 would show two monophyletic groups. *Paraphyly* means that a group arises from one point within another group, and thus the stem group does not contain all of its descendants. In Figure 4.1, population A is paraphyletic relative to population B, since all B genotypes arise from a branch joining them with A2. *Polyphyly* means that individuals within a group arise from various different branches which also give rise to individuals of other groups; in other words, subsets of

FIGURE 4. Tree length distributions and parsimony trees for exhaustive search of genotypes in populations A and B (Fig. 4.1), A and C (Fig. 4.2), and B and C (Fig. 4.3). An arrow indicates the length of the shortest tree, which is compared to the mean and the standard deviation of all possible trees. A negative g_1-statistic indicates the

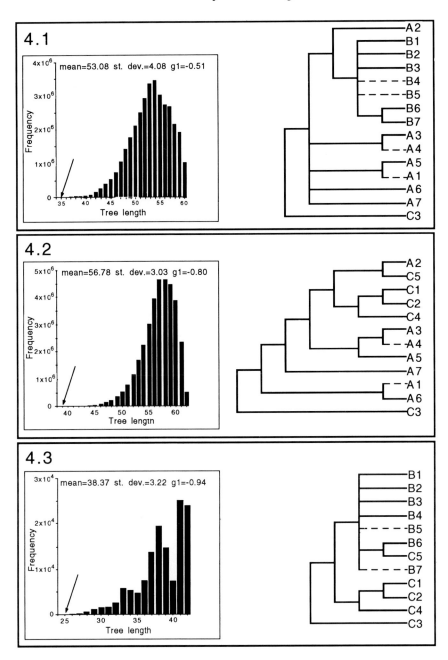

degree of skew in the length distribution. Strict consensus trees are presented when more than one tree has the minimum length (Fig. 4.1 summarizes 30 trees; Fig. 4.3 summarizes 5 trees). In order to limit the number of genotypes analyzed to 10 or 11, genotypes which differed by one unique (unshared) difference only were removed from the search and then added to the tree with a dashed line adjacent to the genotypes from which they differ by one step.

individuals within the group are not in sister groups, and thus the most recent common ancestor is outside the group.

If one panmictic population were separated into two, each of these would initially be polyphyletic, with time progress through paraphyly, and eventually become monophyletic (Avise et al. 1987). Monophyly can be taken as evidence of genetic isolation (Hey 1991; Slatkin and Maddison 1989). Paraphyly, particularly in conjunction with fixed differences leading to the derived group, can also be taken as evidence of genetic isolation (Hey 1991; Fig. 4.1). Polyphyly is equivocal since populations may be reproductively isolated but not have been separated long enough to show monophyly (Fig. 4.3). In general, poly- phyly suggests genetic exchange. Thus we can conclude that both populations B and C have arisen from genotypes within population A (Figs. 4.1 and 4.2, respectively), and that all populations except A and B are, or recently were, exchanging genes. Below we will estimate the actual levels of gene flow between populations.

3.2.4 Cladistic Estimate of Gene Flow

A phylogeny of alleles in one locus can be extremely useful for estimating the level of gene flow between populations (cf. Slatkin and Maddison 1989). Gene flow is normally calculated as the number of immigrants per generation (m). This value is difficult to calculate; thus genetic exchange between popula- tions is normally expressed as $N_e m$, or the effective population size times the number of immigrants per generation. An important value is $Nm = 1$, below which genetic drift between populations becomes highly probable (Slatkin 1985). Genetic exchange at levels of $Nm > 1$ is sufficient to keep populations from diverging unless selection is extremely strong (ibid.).

Allele frequency data from several polymorphic loci, as well as private allele methods, have been used extensively to calculate F_{st} values from which one can estimate Nm (reviewed in Slatkin 1985). Slatkin and Maddison (1989) present a method based on phylogenies of alleles which is as useful as an F_{st} calculated from 10 protein loci (Slatkin 1991) and avoids problems Lewontin (1985) attributes to estimates of Nm based on F_{st}. Their method assumes that populations follow the island model, and that the true phylogeny for a nonrecombinant DNA fragment is known. The phylogeny is used to determine the minimum number of migration events required to explain the geographic distribution of alleles. A migration event(s) is postulated when we observe a discordance between the geographic and phylogenetic distribution of an allele: the geographic location is treated as a multistate unordered character, and each polymorphic state in a branch tip or internal node is assigned a value of $s = 1$. The sum of migration events (sum of s values) is compared to simulation results presented by Slatkin and Maddison (1989) in order to estimate Nm.

In Figure 4.1 observe that only one node (the one joining the B clade to genotype A2) is polymorphic when comparing populations A and B. An $s = 1$

yields a minimum estimate of $Nm < 0.1$, with a standard deviation of less than 0.4. A comparison between populations A and C (Fig. 4.2) shows $s = 3$ and yields $Nm = 2$, with a standard deviation of about 1. Finally, if we compare populations B and C (Fig. 4.3), we find one polymorphic node and one polymorphic branch tip, which together yield $s = 2$ with a minimum estimate of $Nm = 0.66$ and a standard deviation of about 0.7. Thus we can conclude that these populations are held together by variable amounts of gene flow, with divergence due to genetic drift being most likely in population B. Drift is merely an expression for chance fixation of genotypes in a population due to the effects of random sampling of genotypes from one generation to the next. Either drift or selection or both can be the forces which take a population from polyphyly through paraphyly and ultimately to monophyly. Yet even with an Nm of 0.1 (between A and B), we can only be about 65% certain that all of the genotypes found in each population have arisen from a common ancestor within each population, and not from immigrant genotypes (Slatkin 1989).

3.3 Molecular Clock, Rate of Divergence, and Mutation Rate

The possibility of adding a temporal scale to molecular evolution is extremely intriguing. Used with caution and consideration, a well-calibrated molecular clock may be quite useful. To begin with, we could estimate the time of divergence between lineages. The time scale could then be used to estimate rates of evolution for other features, thus adding a temporal element to comparative biology. Finally, if mutations are neutral or have a selection coefficient of much less than the absolute value of $1/N$, then the substitution rate can be used to estimate the mutation rate (Kimura 1983, p. 47).

Due to degeneracy in the genetic code, genes evolve via mutations affecting amino acid replacement as well as through silent mutations. Silent mutations occur at a much faster rate. As a consequence, it makes little sense to consider a molecular clock based on an overall substitution rate. It may be better to consider the molecular "protein clock" proposed by Zuckerkandl and Pauling (Zuckerkandl 1987) separately from the "neutral clock" proposed by Kimura (1983, p. 100). The Zuckerkandl model is based on the observation that amino acid changes tend to accumulate linearly over time. The Kimura model is based on the conclusion that most silent mutations should be neutral or only slightly deleterious, and thus their accumulation should reflect a fairly constant mutation rate. Factors which cause excessive variation in substitution rates have been discussed by several authors (Kimura 1983; Ohta 1987; Dover 1987; Takahata 1987; Palumbi 1989; Gillespie 1991). The existence of a clock per se is not debated as much as is its accuracy and the statistical distribution of the underlying process. Theoretically, any gene can be used as a clock, so long as its mode of evolution (i.e., degree of constraint and distribution of variation) is understood.

As implied, the Zuckerkandl model is suitable for protein coding genes whereas the Kimura model is best for silent positions in protein coding genes, introns, or structural genes. In order to apply a molecular clock to non protein coding regions, one needs to be fairly confident that most mutations are neutral. This condition is most likely met in rapidly evolving, relatively unconstrained regions, such as introns of nuclear genes or the control region of animal mitochondrial DNA. For the Kimura model to be applicable, fairly similar sequences must be compared. Minimally, one must correct the observed difference for multiple mutations (i.e., corrected divergence; Nei 1987, p. 64). In the worst case it may become difficult even to align the sequences. Molecular clocks assume that all positions evolve at the same rate, or more specifically, a Poisson distributed rate. If the substitution process deviates significantly from this distribution, it will be difficult to apply the proper correction. There is evidence, at least for humans (Kocher and Wilson 1991), that the d-loop does not follow a Poisson distribution. Errors in time estimates based on d-loops would therefore increase at higher levels of divergence.

Calibration of a molecular clock is usually accomplished through the use of well-dated fossil material. For example, the split between the family which contains kangaroo rats and the family containing pocket gophers is dated at 20–23 million years ago (Wood 1935; Hafner and Hafner 1983). A comparison of replacement changes in a 432-bp fragment of cytochrome b between two pocket gophers and 16 kangaroo rats (Villablanca and Thomas 1993) yielded an average divergence of 5.48% (st. error 0.32). This corresponds to an average rate of 0.27%/MY. Irwin et al. (1991) estimated that the replacement rate of the entire cytochrome b gene in mammals is 0.36%/MY. These two estimates are compatible since the 432-bp fragment is from one of the more conserved regions of the gene (ibid.).

The rate of 0.27%/MY can be used to estimate the time since divergence between species of kangaroo rats. These times can then be used to calibrate the rate of substitution in the more rapidly evolving control region. The observed difference between 24 pairs of kangaroo rat control regions, corrected using Kimura's two-parameter model (eq. 5.5 of Nei 1987) and divided by the time since divergence for each pair of species, yields a mean divergence rate of 3.0%/MY (st. dev. 0.74) for the 421-bp fragment. If we make the reasonable assumption that all control region mutations are neutral, then the rate of substitution per nucleotide site per generation is equal to the mutation rate (Kimura 1983), or 15×10^{-9} (st. dev. 3.6×10^{-9}).

The rate of divergence can be used to infer the history of *D. panamintinus* populations. Considering the historical biogeography of American southwestern deserts plant assemblages, we would conclude that population B has been separated from populations A and C for at most 8,000 to 13,000 years (Spaulding 1990). Yet considering the phylogeny in Figure 1 we obtain a different estimate. All of the B genotypes differ from genotype A2 by three to five mutations (mean 3.9, st. dev. 0.8). This is equivalent to 0.7–1.2% difference, or roughly 200,000 to 400,000 years. Therefore, postglacial vegetational changes (Spaulding 1990) alone cannot be responsible for the degree of differentiation of population B. Given that on average it takes 2 times the female effective population size $(2N_f)$ generations for a population to become monophyletic, in order for genotype C5 (Fig. 4.3) to have been retained from an ancestral population (i.e. para or polyphyly) the mean female population size of B would have been N_f greater than $200,000/2$ to $400,000/2$. Alternatively, if the population size of B were much less than this (i.e. time for monophyly has been sufficient), genotype C5 would have arisen in population B and flowed from B to population C. Unfortunately, the mean time of $2N_f$ has a very high variance. We will address this issue and test the alternative predictions regarding population B in Section 4.

Some aDNA samples are themselves fossils (more specifically, subfossils). As such, they should be useful for direct calibrations of molecular clocks (see Runnegar 1991): ancient samples stop accumulating mutations at the time they are deposited, while extant lineages continue to evolve. Unfortunately one cannot simply compare the percentage difference between two samples and divide by twice the time since divergence (once for each lineage). A phylogenetic approach is more appropriate. Consider a tree constructed from n_j modern sequences and n_i ancient ones, which contains $n_k = (j - 2 + i)$ internal nodes. If we know the time t in million years since the ancient genotype i was deposited, and we can find the node k furthest from the root which unites genotypes i and j, then the mean substitution rate per million years for one direct calibration is given by

$$\frac{\sum(d_{ij} - 2d_{ik})}{n_i n_j} \times \frac{1}{t_i}$$

where d_{ij} and d_{ik} are the number of substitutions/number of sites between genotypes i and j and between genotype i and node k, respectively, and $n_i n_j$ is the number of pairwise comparisons. This estimate is based on the number of changes inferred from the phylogeny. If sequences are so divergent that a correction measure is needed, then it becomes necessary to infer the ancestral sequences at all nodes and use these sequences to estimate the corrected d_{ik}. Likewise, if estimates are available from population size samples, and intrapopulation variation is significant, it will be necessary to estimate a mean d_{ij} and d_{ik} by subtracting the mean variation within populations from the variation between populations (cf. Wilson et al. 1985).

4. Population Genetics

The origin and maintenance of genetic variation is the central problematica of population genetics (Lewontin 1991). Studies of genetic variation at the allozyme level have not been as illuminating as originally hoped (ibid.), nor have DNA studies employing RFLPs (Wilson et al. 1989). DNA sequences, on the other hand, hold great promise (ibid.; Lewontin 1985). For example, we already know that much, if not most, variation at the DNA level is selectively neutral (Kimura 1983); selection is limited to variation that wanders too far from neutrality. Importantly, the competing hypotheses of selection and neutrality are both testable (Kreitman 1991; Tajima 1989a). Population parameters as well help determine the amount of variation: effective population size, gene flow, genetic isolation, and population structure are all significant components. The mutation rate, or rate of origination of variation, is also a parameter which can be estimated under specific assumptions (see above). Altogether, the potential impact of DNA studies on population genetics is enormous. Population genetics, in turn, can contribute tremendously to our understanding of the evolution of demes, populations, species, and higher groupings.

4.1 Genetic Diversity at the Nucleotide Level

Genetic variation is currently best understood in the context of the coalescent (reviewed by Hudson 1990; Birky 1991). Put simply: if it takes on average $2N_f$ generations back in time for two mitochondrial genotypes to coalesce onto one node (ancestor), then the quantity $2N_f u$ (or $4N_e u$ for diploid nuclear genes) measures the expected number of nucleotide differences per site between randomly chosen genotypes, given that u is the mutation rate per nucleotide per generation, and N_f is the female effective population size. The parameter $2N_f u$ is termed *theta*. It is important to note that theta equals $4Nu$ when considering diploid nuclear genomes, while for organelle genomes the value of theta is $2Nu$. Theta (or more exactly its estimator π) is a useful measure of variation in its own right, and with additional assumptions can be quantitatively related to many population parameters. For example, π gives the average internode length and mean coalescence time for genotypes in a population (Hudson 1990), and it can be used to calculate F_{st} values between populations (Slatkin 1991) as well as Nm (gene flow times population size; see above, Slatkin 1991). Regrettably, the coalescence times of alleles in a phylogeny are exponentially distributed, while mutations accumulate following a Poisson or negative binomial distribution. Consequently, π has a very large variance (Tajima 1983). For this reason, alternative approaches which estimate coalescence parameters will be very useful, if and when they become available.

4.1.1 Estimating Theta and Its Variance

Estimation of parameters from the coalescent theory requires explicit assumptions: the process is based on an ideal Wright-Fisher model (discussed in Birky 1991; Hudson 1990). It requires us to explicitly assume neutrality, a molecular clock, equilibrium with respect to mutation and random drift, and that the total tree length equals the actual number of mutations. This last assumption is based on an infinite-sites model (Kimura 1983, p. 46), which states that the probability of multiple mutations at the same nucleotide position is infinitesimal. Invariant positions, rate variations across positions, and saturation are empirical facts which point to the weaknesses of the model. Hudson (1990) presents several alternative models of evolution which exploit the coalescent and allow for processes such as: selection, hitchhiking, and recombination. His work is an excellent review of the utility and power of the coalescent.

Theta can be estimated directly from sequence data using either of two estimators. The first (eq. 10.5 of Nei 1987) uses the observed number of differences (π = nucleotide diversity) to estimate the expected number of differences. The variance of π (eq. 10.9 of Nei 1987, or eq. 16 of Hudson 1990) is a function of the estimated value of theta and the number of sequences examined. The second estimator of theta (Hudson eq. 17 and Nei eq. 10.3) uses the number of polymorphic (segregating) nucleotide sites per sequence (S of Hudson, and p_n of Nei), corrected to be unbiased by sample size.

4.1.2 Nucleotide Diversity in Modern and Museum Populations

Estimates of π, plus or minus one standard error, are presented in Figure 5. Values are obtained from Nei's (1987) equations 10.5 and 10.9. These results show that mean nucleotide diversity within populations is stable over time even though the actual alleles represented in the samples differ from one time point to the next. As a result, museum populations can be used to

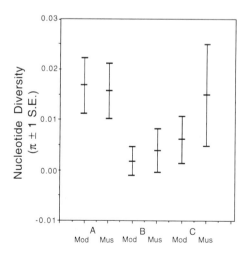

FIGURE 5. Nucleotide diversity ($\pi \pm 1$ st. dev.) among three geographic populations of *D. panamintinus*, comparing modern (Mod) and museum (Mus) samples for each population.

establish base levels of genetic variation in populations. The effect of sample size on confidence limits of π can also be seen in Figure 5. The population with the smallest sample size (Mus C, $n = 8$) is also the one with the largest range, represented by one standard error. The standard errors given are actually underestimates; we are assuming a constant mutation rate across positions, which is clearly unrealistic (Kocher and Wilson 1991).

Based on percentage sequence divergence and a molecular clock we estimated above that population B diverged about 200,000 to 400,000 years ago. It was predicted that either population B contained genotype C5 as a polymorphism retained from the population ancestral to B and C, or that C5 arose within population B and is found in C due to gene flow. The ancestral polymorphism hypothesis predicts $N_f \geqslant 100,000$–200,000, while the gene flow hypothesis predicts $N_f < 100,000$–200,000. Using π and the mutation rate, it is possible to estimate an N_f of 74,500 for population B, which would lead us to reject the ancestral polymorphism hypothesis. Yet, when the variance of π is taken into account, it becomes obvious that it is impossible to say whether the genotype shared between these two populations was present in a population ancestral to both, or whether it arose in one and flowed to the other.

4.1.3 Demonstrating Significant Changes in Nucleotide Diversity over Time

Loss of genetic variability is primarily a function of the effective population size. It is generally assumed that variation is lost at a rate of $\frac{1}{2N_e}$ and $\frac{1}{N_f}$ for diploid and haploid genes, respectively (Wright 1969, p. 212). Thus the only cases in which changes in genetic variation could be observed would be if populations changed in size (ibid.) or if selection acted to change the frequency at an allele and linked loci (Kreitman 1991; Kaplan et al. 1989). Yet even if a population was significantly reduced in size, it would take N generations for it to return to drift-mutation equilibrium. Using temporal samples of aDNA to show changes in genetic variation would require that the time interval (t) between these temporal samples correspond roughly to the new N of the population (i.e. $t = N$) and that the population be held at roughly the same N for the period between samples. As an example, assume a population of $N_e = 1,000$ exists at time $t = 0$, and that the population falls to $N_e = 100$ at time $t = 1$; then a sample from $t = 0$ and $t = 100$ (in generations) would be required in order to estimate the full loss in variability due to the reduction in population size. Samples from any time prior to $t = 100$ would not be as useful, since the population would still be responding to drift.

The large variance of nucleotide diversity further complicates the issue. Fortunately, as the number of alleles (n) in the sample increases, the variance of π decreases (Tajima 1983), though the advantage is hardly noticeable past an n of about 10. It is also noteworthy that the variance of π increases with increasing π (ibid.). To exploit this property of π, one could utilize genes with low mutation rates when comparing populations in the hundreds of thousands. Since π is a product of N_e and u, decreasing u when N_e is held constant will in turn yield a lower estimate of π and thus a lower variance. Using genes with faster mutation rates is preferable for smaller populations: the estimate of π may be less precise since π and therefore its variance are higher, yet the mean will be accurate if (and only if) the locus evolves quickly and variation is observed.

In order to investigate the utility of nucleotide diversity for estimating reductions in genetic variability, the expected π for kangaroo rat d-loops has been projected over a wide range of N_f values. The results (Fig. 6) are obtained for $u = 1.5 \times 10^{-8}$, 500 base pairs, and 10 alleles sampled from 20 individuals. The figure can be interpreted by finding the mean π value up the N_f scale which lies outside of the 95% confidence interval identified from standard errors of π values down the N_f scale. Generally speaking, the larger population would have to differ by a factor of three to four from the smaller population in order for the mean of the larger population to lie outside two standard errors of the mean of the smaller population.

When the purpose is to test for changes in the level of genetic variability over time, it may be more appropriate to use estimates of theta based on the number of segregating sites (S of Hudson, and p_n of Nei) rather than π. Tajima (1989b) has shown that when a population is in a bottleneck, the current population size affects S more than it affects π, while the size of the original population has a greater effect on π than it has on S. Following the bottleneck, S will recover relatively quickly while π will reflect the bottleneck for a much longer time. In contrast to the comparisons of π values over time (Fig. 5) the numbers of segregating sites show pronounced variation between

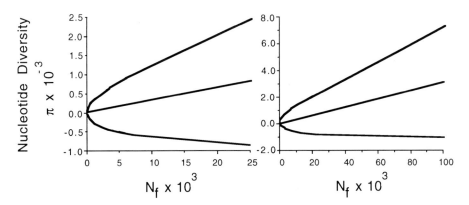

FIGURE 6. Nucleotide diversity (π) as a function of effective population size (N_f). Curves bracketing the central line (mean π) indicate 2 standard errors. The panel on the right is a continuation of the one on the left. Note the difference in scales. See text for discussion.

temporal samples of kangaroo rats. Population A has the most constant S: its value is 10 for both museum and modern samples. Even so, when corrected for sample size (Tajima 1989b), S shows a reduction of 10% over time. Population B shows a decrease of 66%, from 6 to 2. Population C demonstrates a drop of 27%, from 11 to 8. These results clearly illustrate the greater sensitivity of S while suggesting that these populations are not in equilibrium. For a 66% reduction in S (as in population B), a 55% reduction is expected in π (Tajima 1989b); yet Figure 5 shows no such reduction. Following Tajima, the only reasonable interpretation of these results, other than stochastic sampling error, is that π reflects the historical size of the B population, while S more closely reflects the current population size. These results cannot be due to diagenetic processes. If degradation of DNA were a factor, then all museum populations should show a larger number of segregating sites. Population A does not support this notion. Likewise, it could be suspected that even though the S values are corrected for sample size, a bias remains. If this were the case, it would be predicted that modern populations, because of their slightly larger sample sizes, would have larger values of S. The results clearly refute this prediction.

4.2 Changes in Gene Frequency over Time

Another aspect of changing genetic variation over time is the frequency of individual genotypes. Since theta is a function of $2Nu$, it is possible to have different genotypes present in different temporal samples without observing a change in π. Thomas et al. (1990) show a change in the frequency and presence/absence of genotypes across temporal samples. The question is whether this change in frequency is significantly different from what would be expected due to chance alone.

Thomas et al. test for significant frequency shifts by trying to show that the probability of identity by descent between temporal samples is not drastically different from that shown within each temporal sample. Yet this approach does not provide for the estimation of a confidence interval, and thus the interpretation is somewhat subjective. An alternative approach is to test whether the observed frequencies are statistically dependent or independent of the sample time. If the frequency differences are independent of sample origins, then the observed variation can be explained as mere sampling error. A $2 \times n$ contingency table can be applied to the genotype frequencies shown in Figure 7. If the results approach significance, it is necessary to apply a correction for small expected frequencies (Sokal and Rohlf 1981, pp. 735–47). Significance values of the G-test are also shown in Figure 7. None of the tests are significant, though this conclusion requires Williams correction (G_{adj}) in the case of population A. We must conclude that variation in frequencies is simply attributable to sampling error. The possibility does remain that the results are significant in the case of population A; larger samples will be required to test this further.

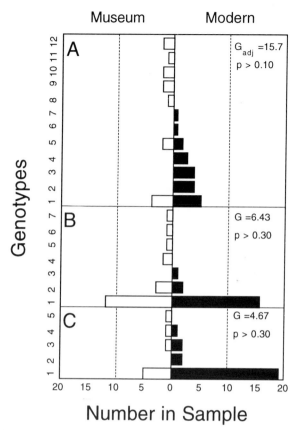

FIGURE 7. Mitochondrial haplotype frequencies for two temporal samples each of three populations of kangaroo rats. Differences in frequency between samples are tested with a $2 \times n$ contingency table. G-statistics and p values are indicated. If the G-statistic is significant, then a G adjusted for small expected values (G_{adj}) is estimated via a Williams correction.

The nature of temporal variation in population A is intriguing. All of the genotypes which are present in the 1911 sample, but absent in the 1988 sample, were found in subadults of the 1911 sample. One possible interpretation is that subadults from neighboring demes dispersed into the deme being sampled, but failed to set up residency and become breeding members of the deme. Though there are alternate interpretations, they all have the same implication: when subadults (roughly equated with nonbreeders) are included in temporal samples, they do not significantly affect the values of π or the G-test, or the conclusions of phylogenetic analysis. Caution might be required in comparing temporal samples taken during different phases of the dispersal cycle in populations where dispersers temporarily increase the number of segregating sites (S) but do not have a high probability of becoming resident breeders.

5. Discussion

5.1 Studies Uniquely Based on aDNA

Many of evolution's most interesting processes have an intrinsically temporal component. Evolutionary analysis of aDNA can afford us the temporal samples needed to study such processes. Yet clearly, the potential applications are still in the process of being identified.

Hybrid zones represent areas of genetic contact between differentiated forms. The processes and dynamics occurring at hybrid zones have a direct impact on speciation and differentiation (Endler 1977; Harrison 1990). Lessa (1992 and pers. comm.) has begun to unlock the temporal information which museum specimens provide. His study investigates a situation in which the population on one side of a hybrid zone has recently come to morphologically resemble the population on the other side. Mitochondrial DNA, in concert with nuclear genes and denaturing gradient gels, allows for the testing of three alternative hypotheses: local extinction with recolonization by a different morph, male-biased gene flow, and morphological convergence without genetic exchange.

Tools to study speciation can also be found in the aDNA tool chest. They offer the potential to directly test long-standing hypotheses regarding the mechanisms and implications of speciation. Some researchers have strongly argued for the importance of the founder flush, or genetic revolution mechanisms, for speciation (Mayr 1954, 1963; Carson 1975, 1982). These hypotheses make explicit predictions regarding the levels and structuring of genetic variation prior to and following speciation. Properly placed temporal samples could provide an avenue for testing such predictions.

Neontological museum specimens have already been applied to the study of evolutionary relationships in extinct species (Thomas et al. 1989). Thomas et al. (1990) and the present study suggest how phylogenetic and population genetic analysis can be extended to rare or endangered species and populations. Additionally, museum collections may contain information which is no longer available from modern populations. One possible example of this is the presence of a shared genotype between museum samples of populations B and C. The genotype (B6 and C5) was found in museum samples with little effort, yet could not be found in modern samples of 19 and 24 individuals. As discussed, this genotype critically affects our interpretation of gene flow between populations B and C.

The possible information contained in archaeological and paleontological specimens is also an area of active research. Several different approaches show promise. The ground-breaking work of Golenberg (cf. Chapter 17) clearly exemplifies two advances. First, as discussed above, given the proper phylogenetic analysis, Miocene samples of *Magnolia* can be used to directly calibrate the molecular clock. Second, these samples lead to specific conclusions regarding the prior geographic distribution of a particular phylogenetic group.

The subfossil material is sufficient to place magnolias in a certain part of North America during the Miocene, while the aDNA from the samples indicates the specific evolutionary lineage involved. Such simultaneous biogeographic and phylogenetic links are usually difficult to establish. The potential of subfossil molecules to reveal the history of organisms and molecules alike is largely unexplored. It seems likely that many interesting studies will be forthcoming in this area.

Phylogenetic reconstructions can be radically affected by the inclusion or exclusion of fossil specimens (Donoghue et al. 1989). There are several reasons for this effect: fossils can identify intermediate and differentiated forms along lineages which later converge; they can resolve short branches along which evolution in rapidly evolving characters can be arrested, so that the record of change is not obliterated; in some cases they are the only hope for resolving the problems of chance similarity between branch tips (cf. Hendy and Penny 1989; Sanderson and Donoghue 1989).

5.2 Biological and Statistical Samples

Successful investigation of many of the issues discussed above will depend on the biological and statistical distribution of the samples. Though it is impossible to foresee all possible issues, some basic guidelines can be provided here.

Phylogenetic analysis can have profound impact on our perception of populations, species, speciation, and comparative biology. Identifying a population as monophyletic, paraphyletic, or polyphyletic is the key, both in terms of gene flow and differentiation, and in determining whether gene trees correspond to population or species trees (Takahata and Slatkin 1990). There is a clear relationship here: the probability of observing paraphyly and polyphyly increases with increasing sample size (ibid.). There is no better argument against the common practice of using no more than one or two individuals per species in phylogenetic analyses. Conclusions at one hierarchical level must be based on adequate sampling of the next lower level. For example, conclusions regarding D. panamintinus can currently be drawn only at the population level, since only individuals within populations have been sampled. In order to draw conclusions regarding subspecies, populations within subspecies would need to be sampled. Similarly, it is clear that in order to make interspecific inferences, one would need to sample populations within species.

Estimates of population parameters may also present particular sampling requirements. Gene flow (Nm) estimates based on phylogenies of alleles are most confidently done from samples of 10 or more alleles (genotypes) per population (Slatkin and Maddison 1989). Since the number of alleles in a population is a function of demographic and historical processes (Slatkin 1987), one is limited to sampling what is available. The number of alleles may be increased by sampling several demes from one population (ibid.).

Confidence in estimates of nucleotide diversity (π) is affected by the number of individuals as well as the number of alleles. The variance of π decreases significantly up to samples of about 10 individuals (Tajima 1983). Above 10 individuals, the improvement is small unless samples are in the neighborhood of 50 individuals (ibid.). Diachronic comparisons of genotype frequency are likewise affected by the number of individuals sampled. Only when samples are large (> 20) can one hope to observe any existing statistical significance. If the expected frequencies of genotypes are low (on the order of 0.05), then a Williams correction is required. This test is conservative (Sokal and Rohlf 1981) and thus requires increased sample sizes. Finally, it seems that diachronic comparisons of genetic variation using π will be most useful when the temporal samples are separated by $t = N \times$ generation time (yrs), where N is the population size following a change in population size. The number of segregating sites (S) may be superior to nucleotide diversity (π) in detecting differences in diachronic comparisons when the number of generations is less than N and when populations are in a bottleneck (Tajima 1989b). Yet if populations are rebounding from a severe bottleneck, then π would be a more sensitive detector of changing levels of genetic variation.

Acknowledgments. This work has been supported by a National Science Foundation Graduate Fellowship and by a grant from the Museum of Vertebrate Zoology. Special thanks are given to Monty Slatkin for helpful discussions and critical review of the manuscript.

References

Aquadro CF, Kaplan N, Risko KJ (1984) An analysis of the dynamics of mammalian mitochondrial DNA sequence evolution. Mol Biol and Evol 5:423–434

Avise JC, Arnold J, Ball RM, Bermingham E, Lamb T, Neigel JE, Reeb CA, Saunders NC (1987) Intraspecific phylogeography: the mitochondrial DNA bridge between population genetics and systematics. Annu Rev Ecol Syst 18:489–521

Bibb M, Van Etten RA, Wright CT, Walberg MW, Clayton DA (1981) Sequence and gene organization of the mouse mitochondrial DNA. Cell 26:167–180

Birky CW Jr (1991) Evolution and population genetics of organelle genes: mechanisms and models. In: Selander RK, Clark AG, Whittam TS (eds) *Evolution at the Molecular Level.* Sunderland Mass.: Sinauer, pp. 112–134

Birky CW Jr, Maruyama T, Fuerst P (1983) An approach to population and evolutionary genetic theory for genes in mitochondria and chloroplasts and some results. Genetics 103:513–527.

Birky CW Jr, Fuerst P, Maruyama T (1989) Organelle gene diversity under migration, mutation and drift: equilibrium expectations, approaches to equilibrium, effects of heteroplasmic cells, and comparisons to nuclear genes. Genetics 121:613–627

Brown WM (1985) The mitochondrial genome of animals. In: MacIntyre RJ (ed) *Molecular Evolutionary Genetics,* New York: Plenum Press, pp. 95–130

Brown GG, Simpson MV (1982) Novel features of animal mtDNA evolution as shown by sequences of two cytochrome oxidase subunit II genes. Proc Natl Acad Sci USA 79:3246–3250

Brown WM, George M Jr, Wilson AC (1979) Rapid evolution of animal mitochondrial DNA. Proc Natl Acad USA 76:1967–1971

Brown WM, Prager EM, Wang A, Wilson AC (1982) Mitochondrial DNA sequence of primates: tempo and mode of evolution. J Mol Evol 18:225–239

Carpenter JM (1988) Choosing among multiple equally parsimonious cladograms. Cladistics 4:291–296

Carson HL (1975) The genetics of speciation at the diploid level. Am Naturalist 109:83–92

Carson HL (1982) Speciation as a major reorganization of polygenic balances. In: Barigozzi C (ed) *Mechanisms of Speciation*. New York: Alan R. Liss, pp. 411–433

Clegg MT, Learn GH, Golenberg EM (1991) Molecular evolution of chloroplast DNA. In: Selander RK, Clark AG, Whittam TS (eds) *Evolution at the Molecular Level*. Sunderland, Mass.: Sinauer, pp. 135–149

Cracraft J, Helm-Bychowski K (1991) Parsimony and phylogenetic inference using DNA sequences: some methodological strategies. In: Miyamoto MM, Cracraft J (eds) *Phylogenetic Analysis of DNA Sequences*. Oxford: Oxford Univ. Press, pp. 184–220

DeSalle R, Freedman T, Prager EM, Wilson AC (1987) Tempo and mode of sequence evolution in mitochondrial DNA of Hawaiian *Drosophila*. J Mol Evol 26:157–164

Donoghue MJ, Doyle JA, Gauthier J, Kluge AG, Rowe T (1989) The importance of fossils in phylogeny reconstruction. Annu Rev Ecol Syst 20:431–460

Dover G (1987) DNA turnover and the molecular clock. J Mol Evol 26:47–58

Endler JA (1977) *Geographic Variation, Speciation and Clines*. Princeton, N.J.: Princeton Univ. Press

Farris JS (1983) The logical basis of phylogenetic analysis. In: Platnick NI, Funk VA (eds) *Advances in Cladistics*, vol 2. New York: Columbia Univ. Press, pp. 7–36

Felsenstein J (1978) The number of evolutionary trees. Syst Zool 27:27–33

Felsenstein J (1979) Alternative methods of phylogenetic inference and their interrelationship. Syst Zool 28:49–62

Felsenstein J (1983) Parsimony in systematics: biological and statistical issues. Annu Rev Ecol Syst 14:313–333

Felsenstein J (1988) Phylogenies from molecular sequences: inference and reliability. Annu Rev Genet 22:521–565

Felsenstein J (1989) PHYLIP 3.2 Manual. Berkeley, Cal.: Univ. of California Herbarium

Gillespie JH (1991) *The Causes of Molecular Evolution*. Oxford: Oxford Univ. Press

Golenberg EM (1991) Amplification and analysis of Miocene plant fossil DNA. Phil Trans R Soc Lond B 333:419–426

Gyllensten U. Wharton D, Joseffson A, Wilson AC (1992) Paternal inheritance of mitochondrial DNA in mice. Nature (London) 352:255–357

Hafner JC, Hafner MS (1983) Evolutionary relationships of heteromyid rodents. Great Basin Naturalist Memoirs 7:3–29

Harrison RC (1990) Hybrid zones: windows on evolutionary processes. In: Futuyama D, Antonovics J (eds) *Oxford Surveys in Evolutionary Biology*, vol 7. Oxford: Oxford Univ. Press, pp. 69–128

Hendy MD, Penny D (1989) A framework for the quantitative study of evolutionary trees. Syst Zool 38:297–309

Hennig W (1965) Phylogenetic systematics. Annu Rev Entomol 10:97–116

Hey J (1991) The structure of genealogies and the distribution of fixed differences between DNA sequence samples from natural populations. Genetics 128:831–840

Hillis DM (1991) Discriminating between phylogenetic signal and random noise in DNA sequences. In: Miyamoto MM, Cracraft J (eds) *Phylogenetic Analysis of DNA Sequences.* Oxford: Oxford Univ. Press, pp. 278–294

Hillis D, Huelsenbeck J (1992) Signal, noise, and reliability in molecular phylogenetic analysis. Jour Hered 83:189–195

Hudson RR (1990) Gene genealogies and the coalescent process. In: Futuyma D, Antonovics J (eds) *Oxford Surveys in Evolutionary Biology,* vol. 7. Oxford: Oxford Univ. Press, pp. 1–44

Irwin D, Kocher TD, Wilson AC (1991) Evolution of the cytochrome *b* gene of mammals. J Mol Evol 32:128–144

Kaplan NL, Hudson RR, Langley CH (1989) The "hitchhiking effect" revisited. Genetics 123:887–899

Kimura M (1983) *The Neutral Theory of Molecular Evolution.* Cambridge: Cambridge Univ. Press

Kocher TD, Wilson AC (1991) Sequence evolution of mitochondrial DNA in humans and chimpanzees: control region and a protein coding region. In: Osawa S, Honjo T (eds) *Evolution of Life.* Tokyo: Springer-Verlag, pp. 391–413

Kreitman M (1991) Detecting selection at the level of DNA. In: Selander RK, Clark AG, Whittam TS (eds) Evolution at the Molecular Level. Sunderland, Mass.: Sinauer, pp. 204–221

Lessa EP (1992) Rapid surveying of DNA sequence variation in natural populations. Mol Biol Evol 9:323–330

Lewontin RC (1985) Population genetics. In: Greenwood PJ, Harvey PH, Slatkin M (eds) *Evolution: Essays in the Honor of John Maynard Smith.* Cambridge: Cambridge Univ. Press, pp. 3–18

Lewontin RC (1991) Twenty-five years ago in genetics: electrophoresis in the development of evolutionary genetics—milestone or millstone? Genetics 128:657–662

Maddison WP, Donoghue MJ, Maddison DR (1984) Outgroup analysis and parsimony. Syst Zool 33:83–103

Mayr E (1954) Change of genetic environment and evolution. In: Huxley J, Hardy AC, Ford EB (eds) Evolution as a Process. New York: MacMillan, pp. 157–180

Mayr E (1963) *Animal Species and Evolution.* Cambridge, Mass.: Harvard Univ. Press

Miyamoto MM, Boyle SM (1989) The potential importance of mitochondrial DNA sequences to eutherian mammal-phylogeny. In: Fernholm B, Bremer K, Jörnvall H (eds) *The Hierarchy of Life.* Amsterdam: Elsevier, pp. 437–450

Moritz C, Dowling TE, Brown WM (1987) Evolution of mitochondrial DNA: relevance for population biology and systematics. Annu Rev Ecol Syst 18:269–292

Nei M (1987) *Molecular Evolutionary Genetics.* New York: Columbia Univ. Press

Ohta T (1987) Very slight deleterious mutations and the molecular clock. J Mol Evol 26:1–6

Orrego C (1990) Organizing a laboratory for PCR work. In: Innis MA, Gelfand DH, Sninsky JJ, White TJ (eds) *PCR Protocols: A Guide to Methods and Applications.* San Diego: Academic Press, pp. 447–454

Pääbo S (1989) Ancient DNA: extraction, characterization, molecular cloning, and enzymatic amplification. Proc Natl Acad Sci USA 86:1939–1943

Pääbo S (1990) Amplifying ancient DNA. In: Innis MA, Gelfand DH, Sninsky JJ,

White TJ (eds) *PCR Protocols: A Guide to Methods and Applications*. San Diego: Academic Press, pp. 159–166

Pääbo S, Wilson AC (1988) Polymerase chain reaction reveals cloning artefacts. Nature (London) 334:387–388

Pääbo S, Irwin DM, Wilson AC (1990) DNA damage promotes jumping between templates during enzymatic amplification. J Biol Chem 265:4718–4721

Palumbi SR (1989) Rates of molecular evolution and the number of nucleotide positions free to vary. J Mol Evol 29:180–187

Pamilo P, Nei M (1988) Relationship between gene trees and species trees. Mol Biol Evol 5:568–583

Runnegar B (1991) Nucleic acid and protein clocks. Phil Trans R Soc Lond B 333:391–397

Sanderson MJ, Donoghue DJ (1989) Patterns of variation in levels of homoplasy. Evolution 43:1781–1795

Slatkin M (1985) Gene flow in natural populations. Annu Rev Ecol Syst 16:393–430

Slatkin M (1987) The average number of sites separating DNA sequences drawn from a subdivided population. Theor Popul Biol 32:42–49

Slatkin M (1989) Detecting small amounts of gene flow from phylogenies of alleles. Genetics 121:609–612

Slatkin M (1991) Inbreeding coefficients and coalescence times. Genet Res Camb 58:167–175

Slatkin M, Maddison WP (1989) A cladistic measure of gene flow inferred from the phylogenies of alleles. Genetics 123:602–613

Sokal RR, Rohlf FJ (1981) *Biometry*, 2nd ed. New York: Freeman, W.H.

Spaulding WG (1990) Vegetational and climatic development of the Mojave Desert: the last glacial maximum to the present. In: Betancourt JL, Van Devender TR, Martin PS (eds) *Packrat Middens*, Tucson: Univ. of Arizona Press, pp. 166–199

Swofford DL (1989) Phylogenetic analysis using parsimony (ver. 3.0). Illinois Natural History Survey

Swofford DL, Olsen GJ (1990) Phylogeny reconstruction. In: Hillis DM, Moritz C (eds) *Molecular Systematics*. Sunderland Mass: Sinauer, pp. 411–501

Tajima F (1983) Evolutionary relationship of DNA sequences in a finite population. Genetics 105:437–460

Tajima F (1989a) Statistical method for testing the neutral mutation hypothesis of DNA polymorphism. Genetics 123:585–595

Tajima F (1989b) The effect of change in population size on DNA polymorphism. Genetics 123:597–601

Takahata N (1987) On the overdispersed molecular clock. Genetics 116:169–179

Takahata N, Slatkin M (1990) Genealogy of neutral genes in two partially isolated populations. Theor Popul Biol 38:331–350

Thomas RH, Schaffner W, Wilson AC, Pääbo S (1989) DNA phylogeny of the extinct marsupial wolf. Nature (London) 340:465–467

Thomas WK, Pääbo S, Villablanca FX, Wilson AC (1990) Spacial and temporal continuity of kangaroo rat populations shown by sequencing mitochondrial DNA from museum specimens. J Mol Evol 31:101–112

Villablanca FX, Thomas WK (1993) Empirical limits to the phylogenetic utility of DNA sequences. M.S. submitted

Walsh SP, Metzger DA, Higuchi R (1991) Chelex 100 as a medium for simple extraction of DNA for PCR-based typing from forensic material. BioTechniques 10:506–513

Watrous LE, Wheeler QD (1981) The outgroup comparison method of character analysis. Syst Zool 30:1–11

Wiley EO (1981) *Phylogenetics: The Theory and Practice of Phylogenetic Systematics.* New York: Wiley

Wilson AC, Cann RL, Carr SM, George M, Gyllensten UR, Helm-Bychowski KM, Higuchi RG, Palumbi SR, Prager EM, Sage RD, Stoneking M (1985) Mitochondrial DNA and two perspectives on evolutionary genetics. Biol J Linn Soc 26:375–400

Wilson AC, Zimmer EA, Prager EM, Kocher TD (1989) Restriction mapping in the molecular systematics of mammals: a retrospective salute. In: Fernholm B, Bremer K, Jörnvall H (eds) *The Hierarchy of Life.* Amsterdam: Elsevier, pp. 407–419

Wolfe KH, Li W-H, Sharp PM (1987) Rates of nucleotide substitution vary greatly among plant mitochondrial, chloroplast, and nuclear DNAs. Proc Natl Acad Sci USA 84:9054–9058

Wood AE (1935) Evolution and relationships of the heteromyid rodents with new forms from the Tertiary of western North America. Ann Carnegie Mus 24:73–262

Wright S (1969) *Evolution and the Genetics of Populations,* vol 2. Chicago: University of Chicago Press

Zuckerkandl E (1987) On the molecular evolutionary clock. J Mol Evol 26:34–46

Sample Preparation and Analysis

4 General Aspects of Sample Preparation

Susanne Hummel and Bernd Herrmann

This introductory chapter on sample preparation and analysis will give a brief overview of the different types and origins of ancient DNA (aDNA) and the different states of preservation. Independent of the handling of the source material, work on aDNA follows some basic principles of sample treatment, DNA preparation, and DNA analysis which are presented on the assumption that the polymerase chain reaction (PCR) technique will be applied. This is especially true for the discussion of sources and effects of inhibitory and contaminating substances. However, one should be aware that some of the basic difficulties associated with ancient materials may also occur when applying other analytical techniques such as hybridization.

1. Types and Origins of Ancient Nucleic Acids

In living organisms almost all cells contain nucleic acids. Under certain circumstances, the different types of tissue and thus the cells and their contents may remain more or less unaltered by diagenetic effects (cf. Chapter 2 and Chapter 17).

The cells of eukaryotic organisms such as animals, plants, and fungi possess different types of nucleic acids. The nucleus of a eukaryotic cell contains linear double-stranded DNA which is condensed by histones within the chromosomes, with genes as subunits. The chromosomes and therefore the chromosomal DNA is inherited from both parents. The nucleus also contains RNA, which is single-stranded and involved in protein synthesis. The mitochondria of the eukaryotic cell, which are responsible for the energy metabolism of the organism, contain DNA, which is almost exclusively inherited maternally. Prokaryotic cells such as bacteria and yeast contain circular DNA and RNA in their cytoplasmic compartment. Viruses finally contain a great variety of nucleic acids. They can consist of either double- or single-stranded DNA, which may be linear or circular and is packed into a protein capsid. Some types of viruses consist of RNA only, which may be double stranded.

Both DNA and RNA show nucleotides as subunits, which are joined together by phosphodiester linkages to form the macromolecule. The nucleotides in turn are composed of a base (adenine, cytosine, guanine, thymine, or uracil), a sugar molecule, and a phosphate group.

2. States of Preservation of aDNA

Immediately after death autolysis sets in, i.e., the organism's own enzymes start to decompose the organic substances of the body. Usually this first phase of decomposition is followed by the decay of soft tissues under involvement of microorganism and fungi and also at an advanced stage, of higher organisms such as insects and vertebrates. This process regularly leads to the complete disappearance of soft tissues; only the mainly inorganic exo- or endo-skeletons and keratinic appendices like hair and feathers (cf. Chapter 15) remain. However, the endoskeletal remains at least from vertebrates are known not to lose their organic components. Besides collagen, fatty acids, etc., there are still cellular structures left, as histological cross-sections of bones show (cf. Chapter 3). In contrast to soft tissues, the organic structures in bones and teeth persist even under normal burial conditions (cf. Chapter 13). This can be explained by the comparatively low water and enzyme content of hard tissues. Each osteocyte or cementoblast seems to undergo a process of individual mummification. Moreover, these cells are protected by their topology against physical and biochemical decay by microorganisms: situated in little caves, they are completely surrounded by protective hard tissue.

A special case of long-term preserved genetic information are plant seeds (cf. Chapter 16). When stored under proper conditions—for example, in the clay walls of ancient buildings—they keep their ability to sprout, in some instances for centuries.

Another form of individual mummification of single cells is observed in dried body fluids, such as evidence blood stains, sperm, or saliva (cf. Chapter 9). In these cases the chance for DNA to persist occurs when the fluid spreads, favoring evaporation.

The destruction of soft tissues may be slowed down or interrupted at any stage of decomposition by either environmental factors or artificial treatment. Examples for conservation of a body by environmental factors are natural mummies and bog bodies. Natural mummification, as can be observed, for example, in most South American mummies, takes place when decomposition due to enzymatic autolysis and decay due to physical and chemical factors are arrested by rapid loss of humidity. This may happen in dry, hot, and draughty climates (cf. section on Dried Samples: Soft Tissues), but also in saline environments. Herbarium specimens may be considered naturally mummified in a broader sense, since no artificial fixatives have been applied (cf. section on Dried Samples: Soft Tissues).

On the other hand, humidity itself may cause conservation of organic tissues if it is accompanied by anaerobic conditions, as in some tombs. One of the most famous tomb burials is the Chinese princess of the Western Han dynasty in Mawangtui (cf. Chapter 1). Medieval crypt burials occasionally reveal partially conserved soft tissues. Humidity combined with anaerobic conditions and a high concentration of humic acids may lead to complete conservation of the body, as demonstrated by bog bodies (cf. Chapter 7). The phenolic properties of humic acids immediately stop the autolytic process by inactivating enzymes. Latrines and cesspits from the Middle Ages often reveal a high variety of well-preserved plant, animal, and human tissues as well (Herrmann 1985).

Examples of decomposition and decay interrupted by intentional treatment with chemical substances like asphalt, natron, alcohol, or formaldehyde are found in Egyptian human and animal mummies, in fixed medical histopathological or organ preparations—either embedded or wet (cf. Chapter 5), and in preserved museum specimens (cf. Chapter 10). A kind of natural fixation and embedding is seen in insects preserved in amber (cf. Chapter 6).

Enzymatic and bacterial activity are also slowed down remarkably by freezing. In natural environments, this condition is found in permafrost soils of arctic regions (cf. Chapter 8) and within glaciers. The most recent and certainly the most spectacular find is the Similaun glacier man of the Ötztal (Austria/Italy) with his excellent physical preservation, although he was probably mummified already when his corpse got into the glacier. The most numerous finds from permafrost and/or anaerobic wet environments are the mammoth remains in Siberia and the famous tombs from Kurgans of the Pazyryk.

The presence of preserved organic substances in fossils (cf. Chapter 17) is at least surprising. The remainder of aDNA in those specimens can only be explained by incomplete replacement of organic by inorganic molecules. Where this occurred, we must suppose that organic matter was "locked" within or between truly fossilized material or sediment layers, which might also have had a protective function in long-term preservation, comparable to the situation in bone.

Since all these types of tissue preservation involve biochemical and physicochemical factors of their own, the state of preservation of organic molecules will be different for each. In order to analyze DNA, the methods of sample pretreatment and the requirements for DNA extraction and purification will vary, depending on the specific qualities of the source material. In principle, there is no difference between, say, extracting DNA from the skin of an animal or a human being. However, there are differences depending on, for example, whether the skin has been wet-preserved or naturally mummified. Accordingly, subsequent chapters will be ordered by type of preservation and tissue quality rather than by biological origin.

3. Specific Obstacles in Work with aDNA

Although decomposition and decay of a body may be reduced or even arrested at an early stage, ancient tissues are always more or less degraded and damaged (cf. Chapters 1, 17). Moreover, the vast majority of specimens do not derive from sterile places; they derive from natural environments like soils and, in the case of museum collections, may have been touched by previous investigators.

For aDNA analysis these circumstances cause special problems:

(a) There may be (modern) DNA contaminations. Since PCR is a very sensitive amplification technique, such contaminations may be sufficient to cause false-positive results.
(b) The average fragment length of aDNA is usually no more than a few hundred base pairs, due to diagenetic effects. There will only be a few intact molecules to serve as templates for PCR, even in well-preserved specimens.
(c) There may be co-purifying inhibiting substances. In PCR analysis they may cause complete inhibition of the enzyme *Taq* polymerase and thus false-negative results.

The basic prerequisites for work on aDNA are therefore:

(a) Techniques for sampling and/or pretreatment of the samples that avoid contamination.
(b) The application of gentle extraction techniques in order not to cause any additional DNA damage. If DNA repair techniques become more valuable for aDNA, these might be performed prior to PCR.
(c) Gentle but stringent purification of DNA extracts to remove any inhibiting components.

Although the following general recommendations are basically aimed at work with PCR, the principal aspects of aDNA preparation, inhibition, and contamination would be the same for the application of any other hybridization technique.

4. Contaminants Causing False-Positive Results

4.1 Sources

Modern DNA may be introduced at any stage of sample processing. It may have come in contact with the sample during excavation or earlier investigations. It may also be introduced during the preparation by contaminated reagents or traces of saliva, skin scales, etc. If nonhuman remains are investigated for species-specific genetic properties, the introduction of whole genomic modern human DNA will usually be a minor risk.

Modern or aDNA of other species may be introduced if the sample is invaded by fungi or bacteria, or if insects or their larvae adhere to the sample. If the primers used in PCR are specific and the amplification procedure is stringent, the risk of false positives should be very small. But the investigator should be aware of possible mismatch sites on the contaminating DNA.

Product carry-over is the highest risk of contaminating a sample. A carry-over contamination results from the transfer of amplification products of former PCR runs to newly set-up reactions. It can occur through any direct contact (via gloves etc.) or even indirect contact by aerosoles (via pistons of pipettes etc.).

4.2 Monitoring for False Positives

There are three types of controls to identify false positives. *Negative controls* are the most universal, since they run through the whole procedure under essentially the same conditions as the investigated samples. A negative control should be as similar as possible to the sample, except for the very genetic property under investigation. It must have similar physicochemical properties and, if at all possible, should be in a similar state of preservation. For example, in amplifying human-specific aDNA fragments from bone extracts one would best choose an animal bone from the same burial site as negative control.

If ancient negative controls similar to the sample are not available, a modern negative control in combination with a modern *positive control* will do the job (physicochemical properties may vary strongly between ancient and modern samples). In this case it is sensible to dilute the modern control to approximate the similar aDNA concentration.

A *blind control* undergoes all extraction, purification, and amplification procedures, except that it does not contain any sample material. This type of control therefore will reveal even minor contaminations in any of the reagents and chemicals used in the procedures.

A *no-template control* serves only for the amplification reagents and conditions. It contains every amplification reagent except DNA.

If contamination is present indeed different types of controls will help to identify the source more quickly than a negative control alone.

4.3 Avoiding Contamination and False-Positive Results

There are various strategies to avoid contamination. To prevent product carry-over, complete physical separation of pre- and post-PCR work spaces is necessary; technical devices should never be exchanged between these areas (cf. Chapter 3). To prevent modern contamination, outer surfaces of specimens should be avoided in sampling or be effectively decontaminated (see below). Samples should always be kept sterile and separate from each other (to avoid cross-contamination) and never be touched without sterile disposable gloves. Sample screening by histological sectioning and staining can be very valuable (cf. Chapter 3) to identify samples that are highly contaminated by microorganisms.

5. Inhibitors Causing False-Negative Results

5.1 Sources

Inhibiting substances may be introduced to the DNA extraction reaction by components of the specimens themselves. Since ancient specimens usually have either been treated with conserving substances or been surrounded by e.g., soil, their chemical components will differ from those of fresh specimens. The asphalt that is part of the Egyptian mummification technique, as well as the humic acids contained in all types of soil, contain phenolic compounds and other potential inhibitors. Since most of these substances are not effectively excluded by the usual extraction techniques, they get into the amplification vessel and inactivate the *Taq* polymerase.

Another source of inhibitors may be a high amount of DNA stemming from origins other than the DNA to be investigated. In the case of a burial, these are mostly microorganisms—bacteria and fungi—which invade bone or remaining soft tissues. When searching for viral components in old embedded histopathological specimens, a high proportion of the DNA content may be human DNA of the sectioned tissue. Besides acting as potentially false-positive contaminants, large amounts of nontemplate DNA may also effect inhibition, by causing competitive reactions during amplification if the non-template DNA contains primer mismatch sites. In the worst case, this may lead to complete inhibition of the specific reaction while nonspecific products are still being generated.

When working with elevated molarities in the extraction buffer, as is necessary in hard tissue preparations, usually the final DNA extract will still contain high salt concentrations. This will affect the sensitive balance of PCR buffers and thus the amplification result.

5.2 Monitoring for False Negatives

Only positive controls are applicable here. Like negative controls, a positive control should show the same, or similar, physicochemical properties as the sample concerning the type of tissue and the state of preservation. But in contrast with a negative control, it must contain the specific genetic information being investigated. Beside an ancient positive control derived for example, from an earlier investigation, one should also run a modern tissue positive control. A modern control allows to exclude positive results exclusively due to ancient-material-specific DNA contaminations from microorganisms, fungi, insects, etc., which may cause amplification in the ancient positive control.

5.3 Avoiding Inhibition and False-Negative Results

There are two simple but nevertheless effective ways to overcome inhibition: extensive DNA purification and, in case of large amounts of nontemplate DNA, DNA separation. The most common purification and separation techniques are referred to in the next chapter. Details on handling can be found in Chapters 5 to 17.

One of the two other-methods commonly used to overcome inhibition is the dilution of DNA samples prior to PCR. This reduces the absolute content of inhibitor and thus its inactivating effect on the polymerase. However, since the DNA content is reduced proportionately, if there are very few templates this may be counterproductive.

The alternative is to add bovine serum albumin (BSA) to the amplification reaction, which has a positive effect on enzyme activity. But of course, the additional reagents with their complex molecular structures may be sources of additional contamination as well. Our recommendation therefore stresses the application of extensive purification and separation techniques in order to overcome inhibition in aDNA amplifications.

6. Principal Steps of DNA Preparation

6.1 Sampling

In working on human remains, specimens which have not been touched by either excavators, preparators, or previous investigators should be preferred to lower the risk of contamination with modern human DNA. Unfortunately, this guideline is not easy to follow, especially when sampling from museum collections. The situation becomes less critical when sampling from animal or plant remains. But it still depends on the genetic property of interest whether contaminating human DNA may bias the result, and it might always act as an inhibitory factor in PCR. To facilitate sampling for molecular biological investigations, it should become the rule to touch specimens with disposable gloves only and to separate samples immediately after excavation, before preparation.

6.2 Sample Pretreatment

Since samples probably have been touched, the outer surfaces of a specimen should be avoided. If this cannot be done, for example because of the small size of a specimen, it is necessary to decontaminate the sample properly. Shortwave UV light (254 nm) is convenient for this purpose: it induces covalent binding of opposite thymine bases, preventing double-stranded DNA from melting; thus these fragments are no longer accessible for amplification. All outer surfaces of the samples should be exposed to UV light for a few minutes.

In the case of hard tissue samples, short exposure to acid can be an alternative. But there is the risk that the acid might destroy endogenous DNA by penetrating the specimen. After decontamination hard tissues are crushed and ground to facilitate extraction by increasing surface area. Liquid nitrogen may be used before grinding to render the material brittle. Soft tissues are often rehydrated before homogenizing, which is often part of the extraction itself already.

6.3 DNA Extraction

DNA extraction serves to release nucleic acids from the cells or cell remains into solution. A typical extraction buffer will contain Tris-HCl, EDTA to break up the cell walls by chelating with calcium ions and to stabilize the DNA, a sodium or potassium salt creating an isotonic milieu to stabilize free nucleic acids, proteinase K to digest the protein fraction, and optional, a nonionic detergent to stabilize buffer and DNA (cf. Chapters 5 to 17 for detailed protocols).

For soft tissues, extraction is usually a one-step procedure taking only a couple of hours. In hard tissues like bone, the mineral barriers must be broken down first in order to gain access to the bone cells. Therefore the extraction medium must contain a high concentration of EDTA, most of which will chelate with the calcium of the apatite mineral. This process may take up to a week. A proteinase K digestion is then performed to free the nucleic acids from adhering proteins.

If a tissue contains a comparatively small protein fraction, or if cells can be removed directly from the tissue into an extraction buffer, it may be sufficient to perform a short boiling procedure optionally supported by a protein digest. In these cases, where only minor cell debris is expected, further DNA isolation steps may be skipped. However, before submitting a crude cell lysate to PCR, dilution or an additional purification step may be necessary.

6.4 DNA Isolation

To isolate DNA from cell debris, usually a phenol/chloroform extraction is performed. This step should not be repeated too frequently on a sample, since every treatment causes further DNA damage at least due to mechanical strain. One must be careful not to retain any phenol in the extraction since even mere traces of it will cause complete inhibition of *Taq* polymerase. It is recommended, therefore to leave generous amounts of the aqueous phase behind when separating it from the organic phase.

When working with high-molar EDTA concentrations and pure phenol, the phases may be reversed after centrifugation, i.e., the aqueous phase may be at the bottom and the organic phase on top. Usually this is easily recognized because of the more or less brownish color of the aqueus aDNA phase. Adding some sterile water to lower the specific weight of the aqueous phase, followed by recentrifugation, will solve the problem.

6.5 DNA Concentration

Concentration and purification of DNA is usually done by ethanol precipitation and optional ethanol washing. Adding two to three volumes of ethanol to the DNA solution works well for precipitation. When extraction and isolation have been carried out with regular buffers, i.e., comparatively low molarities, the addition of high-molar salt solutions may help the precipitation process. This is not necessary, and may even hinder the process, when working with high-molar buffers, since EDTA salts have a precipitating effect already.

An alternative to concentration and purification is the centrifuge-driven mechanical separation of DNA from the aqueous solution, as for example with a porous-membrane device.

6.6 Additional Purification

The isolation and concentration of DNA from ancient samples often does not result in purification. Further purification can be achieved by selective binding of the DNA to a resin or glassmilk, washing of this DNA complex, and subsequent elution of the low highly purified DNA. This technique may be performed before or right after phenol/chloroform extraction. If the DNA binds well to the resin or glass beads, even in crude lysates, the method avoids a lot of DNA loss and further damage that would be incurred during multiple phenol/chloroform treatments.

An alternative to resin or glassmilk binding is the electrophoretic separation of DNA from other co-extracted and isolated compounds. Since salts and substances derived from soil or other humic acid containing sources reveal green-blue fluorescences on an ethidium bromide stained agarose gel, the orange fluorescent DNA portion may be selectively cut out, melted, and re-extracted.

Examples for the application of all of these methods, specific handling suggestions, reagent concentrations, and references to the relevant literature will be given in the chapters below. A good introduction to procedures and devices is found in Sambrook et al. (1989). Reagent kits are accompanied by instructions as well.

References

Sambrook J, Fritsch EF, Maniatis T (1989) *Molecular Cloning: A Laboratory Manual*, 2nd ed. Cold Spring Harbor, N.Y.: CSHL Press

Herrmann B (1985) Parasitologisch-epidemiologische Auswerungen mittelalterlicher Kloaken. Zeitschrift für Archäologie des Mittelalters 13:131–161

Fixed and Embedded Samples
5 Screening for Pathogenic DNA Sequences in Clinically Collected Human Tissues

Wayne W. Grody

1. Introduction

The sources of DNA discussed in this chapter are not nearly as old as those covered elsewhere in this book and, indeed, can hardly be considered "ancient" at all. Yet within the realm of diagnostic clinical medicine, for whose purposes these specimens are most often prepared, they are about as old—with the possible exception of certain extraordinary forensic or archaeological investigations—as any human tissues examined.

Patient tissue specimens, whether obtained at surgery, biopsy, or autopsy, go first to hospital pathology departments, where they are initially examined grossly in the fresh state. Representative or clinically suspicious areas are then dissected out, immersed in chemical fixatives, and embedded in blocks of paraffin wax for sectioning on a microtome. After mounting on glass slides, staining, and microscopic examination of the 5-μ sections, the remainder of the paraffin block, still containing the bulk of the embedded tissue, is stored in the department's surgical pathology or autopsy files. While the bulk of the unembedded original specimen is eventually discarded (usually within a few weeks), the filed paraffin blocks, as well as the stained slides containing the sections cut from them, are retained indefinitely—essentially forever. Initially this practice evolved for obvious medicolegal reasons: should any question later arise regarding the histopathologic diagnosis rendered on a particular specimen, the slides can be pulled for re-examination and additional microscopic sections prepared if necessary.

But at the same time, these samples become, in aggregate, a valuable archival resource of human tissues exhibiting a broad spectrum of pathologic processes, available to clinical investigators for retrospective research. It is the convention of most academic medical centers that clinical specimens, including embedded tissues, be made available for legitimate research purposes once the indicated clinical laboratory testing on them has been completed and reported to the attending physician. Moreover, as long as the research results derived from such specimens are kept anonymous and not tied to a particular patient by name, it is accepted that such research can proceed without obtaining prior informed consent from the patients or approval from institutional human subject protection committees. In that sense, these specimens become analogous, at least for research purposes, to the more ancient and always anonymous sources of human material covered elsewhere in this book. The sole difference is that the investigator must try to avoid, unless absolutely necessary, exhausting the entire amount of embedded tissue in the course of the research; some must be retained for possible future medico-legal investigations, as already mentioned. In addressing this consideration, Radosevich et al. (1991) have developed a nondestructive method for determining the minimum volume of tissue in a paraffin block which must be extracted to yield a sufficient amount of DNA for study. By estimating the number of 5-μ sections needed, based on morphologic tracings of a cross-section, they were able to spare the remainder of the tissue for later use or return to archival storage.

For most of the history of anatomic pathology (that branch of the discipline which deals with the study of disease-related changes in tissues and organs), the assumption in retaining embedded samples, whether for review of clinical diagnosis or for research, was that they would be examined at some later date by essentially the same methods as had been applied when they were first obtained—i.e., by basic histologic observation, perhaps with the addition of some newer histochemical or immunochemical stains, or by electron microscopy. The thrust of the investigation, like that of anatomic pathology as a whole, was always purely morphologic—whether at the gross, light microscopic, or ultrastructural level. It is only within the last few years, coincident with the explosion in molecular biology, that these countless tissue blocks gathering dust in pathology departments all over the world have come to be recognized as a resource for more than just a large series of histologic patterns. With the enhanced insight provided by recombinant DNA technology, we can now see that these aging and mundane little blocks are also the bearers of a vast amount of genetic information. For within them is contained not only the histopathologic record of the patients' diseases and lesions, but also the entire genetic makeup of the patients themselves, their tumors, and any microorganisms that had colonized them.

This expansion of the traditional domain of diagnostic pathology has been nothing short of revolutionary, and has given birth to a new subspecialty within the field. Our newfound capability to examine the nucleic acids in any pathologic specimen, and thus to perform pathologic study at the molecular level, has been designated "molecular pathology"; it has, virtually overnight, thrust pathology from its very traditional place to the outermost frontier of modern medicine (Grody et al. 1989a). It is this domain that is the focus of the present chapter. While many of the same techniques are also available to veterinary pathologists, and there are methods for molecular analysis of embedded plant tissues as well (McFadden et al. 1988), this chapter is concerned exclusively with embedded human tissues, in most cases those obtained in the course of routine clinicopathologic and diagnostic procedures in hospitals and clinics.

1.1 Fixing and Embedding of Human Tissues for Pathologic Examination

In contrast to many of the other sources of DNA discussed in this book, which have undergone natural decay in the environment over the millennia or have by chance remained partially preserved through freezing, immersion, or entrapment in resin, human tissues collected clinically for pathologic diagnosis are treated quickly and deliberately to retain their physical structure and render them suitable for sectioning. Aside from the minority of cases for which a surgeon requests intraoperative biopsy diagnosis of a frozen section (which are not discussed further here since frozen specimens are essentially equivalent to fresh specimens from a biochemical/molecular perspective), such treatments will almost invariably involve immersion in a chemical fixative followed by impregnation and embedding within some solid matrix. These procedures halt autolytic enzyme digestion and bacterial putrefaction, and thus must be performed as soon as possible after removal of biopsy tissue from a living patient or, in the case of tissues obtained at autopsy, as soon as possible after death.

A wide variety of fixatives and embedding procedures have been developed over the long history of anatomic pathology, and a detailed listing of them is beyond the scope of this chapter. However, a few of the most common will be mentioned, ignoring for our purposes any broad consideration of their relative advantages and disadvantages save for their effect on nucleic acids, which will be covered below. The most commonly used general-purpose fixative in anatomic pathology currently is a 10% solution of buffered formalin (formalin itself being a 37% solution of formaldehyde in water). Also used sometimes are Bouin's fixative, containing formalin, picric acid, and acetic acid; and Zenker's fixative, composed of formaldehyde, mercuric chloride, and potassium dichromate (Junqueira and Carneiro 1971). For electron microscopy, a two-step fixation procedure is often used, consisting of 2–6% glutaraldehyde solution followed by osmium tetroxide. These fixatives act primarily by chemically cross-linking and/or precipitating tissue proteins.

The main goal of the embedding process is to render the fixed tissues firm enough for easy sectioning on a microtome and mounting on glass microscope slides. The most common embedding material for routine light microscopy is melted paraffin, which subsequently hardens at room temperature. Tissues are first dehydrated by immersion in successively higher concentrations of ethanol in water (from 70% to 100%) and then infiltrated with xylene. The xylene is then removed by heating to 60°C so that the paraffin can enter. If the fixed tissues have been placed in small plastic cassettes prior to the embedding sequence, the hardened paraffin will assume the shape of a cuboidal block, which can be handled easily and conveniently mounted on the microtome for sectioning. The typical thickness of sections for routine light microscopy as well as the in situ hybridization technique discussed below is 4–5 µ. For the much thinner sections (0.02–0.1 µ) required for electron microscopy, tissues are typically embedded in plastic resin (such as Epon) rather than paraffin (Junqueria and Carneiro 1971).

Not all anatomic pathology examinations focus on sections of solid tissues. In the field of cytopathology, slides composed of dispersed single cells, obtained by natural exfoliation into body fluids or by fine-needle aspiration of a solid tissue or by simple smearing of a drop of blood, are the targets of diagnostic investigation. Such specimens may simply be air-dried on the glass slides, or fixed with formalin, Papanicoulou fixative, or some other reagent. While not technically "embedded" and therefore somewhat less germane to this chapter, they are nevertheless equally valid, and sometimes superior, substrates for molecular biological analysis as tissue sections.

1.2 The New Discipline of Diagnostic Molecular Pathology

The newest and perhaps most revolutionary area of laboratory medicine involves the application of techniques of modern molecular biology to diagnostic clinical testing. This discipline uses sophisticated analytic techniques, formerly confined to the domain of molecular biology research laboratories, to dissect the genome of a patient—or that of the microbes infecting a patient—within a hospital setting. It is in this newborn subspecialty of pathology that some of the most exciting and powerful approaches to human disease are at hand, and also where the traditional boundaries between basic research and clinical medicine are most blurred, leading to some unprecedented ethical and legal dilemmas. Examination of the genetic material (DNA, RNA, or whole chromosomes) in a clinical specimen has applications over the entire spectrum of medicine, including infectious diseases, cancer, and genetic diseases. In each of these areas, the new technology holds great promise not only for accurate diagnosis but also ultimately for specific gene therapy.

The techniques employed in diagnostic molecular pathology laboratories are the same ones used in basic molecular biology research laboratories and indeed in all the studies of ancient DNA discussed in this book: DNA and RNA extraction and purification, cleavage and analysis of DNA with restriction endonucleases, Southern blotting, DNA sequencing, in situ hybridization, and the polymerase chain reaction (PCR). The aim of each of these techniques is to detect the presence of a foreign DNA sequence (i.e., that of an invading microorganism) within the clinical specimen, or to reveal something about the nucleotide sequence of the patient's own DNA. The basic biochemical process underlying all these techniques, and that which confers upon them such great specificity, is nucleic acid hybridization—the specific binding of DNA or RNA molecules through complementary nucleotide base pairing.

In this sense, molecular pathology procedures truly stand apart from those performed elsewhere in the clinical laboratory; for these are the only clinical procedures in which the analytes we are measuring, and the reagents we use to detect them, are the same: they are both nucleic acids. Furthermore, these are the only clinical laboratory reagents and analytes which have the capacity to replicate. But perhaps what most sets these techniques apart and engenders such excitement is the knowledge that every time we use one of them to diagnose disease in an individual patient, we also contribute incrementally to man's broader knowledge of the basic molecular biology of that disease.

1.3 Clinical Use of DNA Probes in Anatomic Pathology

As that subspecialty of pathology dealing with diagnoses made on tissue, it is anatomic pathology which prepares and utilizes embedded specimens. Thus, we will not consider here such unfixed patient samples as blood, urine, cerebrospinal fluid or other body fluids which fall under the domain of clinical pathology.

The various investigative modalities to which formalin-fixed, paraffin-embedded tissues are routinely subjected, including standard light microscopy, special stains, and immunohistochemistry, have already been mentioned. In some sense, DNA hybridization or amplification may be considered just another natural extension of these, though one with its own unique properties, as discussed in the previous section. In fact, it is quite analogous in many ways to immunohistochemistry, in which fixed tissues are probed for the presence of particular antigens by the application of specific polyclonal or monoclonal antibodies and a colorimetric detection system. That technique is employed most often in the differential diagnosis of poorly differentiated malignancies, in which the identification of a tissue-specific protein product may provide some clue as to the site of origin (e.g., detection of insulin might identify a tumor as arising from pancreatic islet cells). Its practical application is often limited, however, by suboptimal specificity of commercially available antibodies, and by destruction or alteration of target antigens brought on by formalin fixation. As described in Section 1.1 above, formalin and other preservatives fix tissues by chemical cross-linking and/or precipitation of endogenous proteins—the same proteins we wish to recognize by antibody probes. Since the epitopes recognized by antibodies may be as small as five amino acids long, it is easy to imagine how even partial damage to a target protein by formalin may drastically reduce the fidelity of antibody binding. It is this phenomenon that is responsible for the notorious problems of false negatives and false positives (otherwise known as "background staining") with which immunohistochemistry is fraught.

In contrast to antigen–antibody methods, probing for nucleic acids in fixed tissue specimens boasts a number of advantages. First, unlike proteins, DNA and RNA are not cross-linked by formalin or other commonly used fixatives, and in fact undergo relatively little direct damage or alteration. The sole chemical effect seems to be the formation of Schiff bases on free amino groups of the nucleotides (Fraenkel-Conrat 1954; Haselkorn and Doty 1960), but this reaction is reversible upon washing in aqueous buffers (Haselkorn and Doty 1960; Jackson 1978). Second, nucleic acid hybridization relies for its fidelity on the specific base-pairing of extensive sequences, often hundreds or thousands of nucleotides in length. Thus, the requirements for a positive reaction are far more stringent than those for the binding of an antibody to a tiny epitope on a protein. Even if related nucleic acid sequences are present in the target tissue, their hybridization can essentially be wiped out by performing the reaction at higher stringency (e.g., raising the temperature or lowering the salt concentration). This concept of variable stringency has no analogy in immunologic procedures. Finally, and perhaps most important, hybridization methods do not require expression of a target sequence into its protein product in order to be detected; the mere presence of the gene itself or its RNA transcript is sufficient. Such considerations become important when screening for genetic disease mutations, genetic alterations in tumors, and latent viral infections.

The (initially surprising) discovery that DNA and RNA remain relatively intact and accessible for hybridization in fixed specimens opened up the heretofore very traditional and unchanging field of anatomic pathology to the most contemporary and rapidly advancing developments of modern molecular biology. It also led to an especially satisfying fusion of the two fields, via the combined molecular/morphologic technique of in situ hybridization. The result is that an increasing number of diagnostic questions that formerly were addressed solely by traditional anatomic pathology methods are becoming amenable to molecular approaches performed on the very same specimens.

The crucial term in the above paragraph, however, is "relatively intact." No one has claimed that the quality of DNA extracted from fixed tissue is equivalent to that from fresh or frozen sources. In fact, its quality can vary enormously, from almost completely intact to almost completely degraded, with the actual degree of degradation depending on a variety of fixation conditions over which the molecular pathology laboratory ultimately receiving the specimens typically has no control. For instance, it has been reported (Tokuda et al. 1990) that immediate formalin fixation at $4°C$ inhibits DNA degradation, compared to fixation at room temperature. Yet few molecular pathology services will be able to prevail upon their surgical pathology colleagues to adopt this nonstandard protocol even prospectively, while it is already too late for archival specimens. Still, a number of inventive methods have been developed which attempt to make the best of a suboptimal situation and produce extractable DNA which is at least partially amenable to genetic analysis. The appropriate methods will vary depending upon the molecular analytic technique to be used: Southern blotting, dot blotting, in situ hybridization, or the polymerase chain reaction.

2. Molecular Methods Applicable to Embedded Pathologic Samples

2.1 Dot Blotting/Slot Blotting

These techniques, requiring only spotting of an extracted DNA specimen on a sheet of nitrocellulose or nylon filter, may be considered the crudest of DNA analysis methods on fixed tissue. Since no sequence-specifc manipulation or nucleic acid sizing is involved, they do not require fully intact target DNA molecules to yield interpretable results. As long as the average DNA target size remains undegraded down to some theoretical hybridization limit, specific probe binding to the spot will occur and a positive signal will be generated. Target DNA that is too fragmented to yield meaningful Southern blot results can still produce interpretable findings on dot blot analysis. Without the need for restriction enzyme digestion and gel electrophoresis, this technique is also appreciably faster than Southern blot.

Yet even this rather simple assay can be problematic when applied to embedded specimens. The quantity of target DNA obtained from such specimens will be significantly lower than what one is accustomed to from using fresh tissues. After only 1 hr of formalin fixation, extractable DNA yields may decrease by a factor of 60–95% (Moerkerk et al. 1990; Warford et al. 1988; Dubeau et al. 1986). And the DNA which is extracted may be difficult to denature into adequate single-stranded hybridization targets (Moerkerk et al. 1990). Also, it goes without saying that any embedded tissue (or portion thereof) that was already autolyzed or necrotic prior to fixation will yield predominantly degraded DNA (Dubeau et al. 1986).

Obviously, this type of analysis reveals little about the genetic fine structure of the material being examined; rather, it is an all-or-nothing type of approach, used to answer the question: Is a particular nucleic acid sequence present in this clinical specimen or not? Consequently, its main applications have been in the realm of infectious disease, in which the foreign DNA sequence of an infecting microorganism is being detected. However, when combined with PCR and allele-specific oligonucleotide probes, it can also be used for detection of single-nucleotide mutations associated with genetic disease (Conner et al. 1983). While embedded samples are hardly the optimal source material for such analysis, they are sometimes the only recourse available when family studies require DNA testing of a previously affected member who is now deceased (see below). The DNA estimation method of Radosevich et al. (1991) mentioned above, which was applied to slot blots, can be used to determine whether sufficient material exists, prior to exhausting the supply unnecessarily. Taking this one step further, Scholefield et al. (1990) developed an alkaline extraction method capable of detecting a single-copy gene by dot blot hybridization of DNA obtained from a single 5-μ section cut on the microtome. Once obtained, dot blots of DNA from fixed material can be successfully reprobed many times (Mark et al. 1987).

2.2 Southern Blotting

Southern blots have for years been the basic workhorse of molecular biology, since their sequence-specific cleavage (by restriction endonuclease digestion) and sizing (by gel electrophoresis) of DNA fragments provide so much information about the genetic makeup of target genomic material. For restriction digest information to be useful, however, the starting DNA must be largely intact (i.e., not degraded or fragmented). The DNA must also be clean (i.e., free of protein) so that restriction endonuclease cleavage sites are exposed and accessible. Prerequisites for successful Southern blotting from embedded tissue are thus quite a bit more rigorous than those required for the other techniques discussed in this chapter—dot blotting, amplification, and in situ hybridization. The tendency for DNA degradation in such samples has already been mentioned, and may be so severe as to render Southern blots uninterpretable (Seto and Yen 1987). And while DNA is not itself cross-linked by formalin, it may become so enmeshed within cross-linked nucleoproteins in a fixed sample as to become relatively resistant to endonuclease digestion (Jackson and Chalkey 1974). In addition, it has been found that formalin can methylate purine nucleotides (Feldman 1973; McGhee and Von Hippel 1977), which could interfere with cleavage by methylation-sensitive endonucleases. Lastly, it has been observed that even completely digested restriction fragments from formalin-fixed DNA may run slower (and thus appear larger) than the equivalent fragments obtained from fresh tissue (MacDonald and Cohen 1990).

For these reasons it was assumed for many years that readable Southern blots could not be produced with DNA extracted from fixed tissue, and such specimens were routinely rejected by diagnostic molecular pathology laboratories. Recently, however, more refined techniques for extracting usable DNA from fixed specimens have been reported, and the relative advantages of various fixatives for this purpose have been evaluated. The first group to do so were Goelz and colleagues (1985). Unlike other workers (Moerkerk et al. 1990; Warford et al. 1988; Dubeau et al. 1986), they found that length of time spent in fixative was not a critical factor, nor was it essential to remove every last trace of paraffin by either heating or dissolving in chloroform or xylene. The crucial step seemed to be fine mechanical mincing of the embedded tissue to allow for adequate proteinase K digestion and detergent (SDS) penetration. In their experience, fragmentation of target DNA was more likely to be the result of increased lag times between surgical excision of tissue and start of fixation than an effect of the fixative itself.

Helpful modifications of these procedures were subsequently introduced by other investigators. Dubeau et al. (1986) found that sequential extractions and differential solubilizations could be used to preferentially remove low molecular weight (i.e., degraded) DNA during sample preparation. Of note, they observed no inhibition of digestion of formalin-fixed DNA by methylation-sensitive endonucleases, despite the theoretical concern mentioned earlier.

Moerkerk et al. (1990) compared the Goelz and Dubeau extraction methods, finding no difference in quality of DNA for Southern blotting between the two. Many have commented that phosphate-buffered formaldehyde does not degrade DNA, whereas unbuffered formaldehyde does (Nuovo and Silverstein 1988; Tokuda et al. 1990). Since the formalin used in surgical pathology laboratories is a buffered solution of formaldehyde, this should not be a concern in routine practice.

Numerous studies have compared DNA extraction yields between formalin and less standard fixatives. The consensus seems to be that ethanol is the best fixative if Southern blotting is to be performed, with results virtually indistinguishable from those obtained with DNA from fresh or frozen tissue (Bramwell and Burns 1988; Wu et al. 1990; Hsu et al. 1991; Smith et al. 1987). Ethanol is also especially good as a transport medium, and in fact the first pilot proficiency tests for diagnostic molecular pathology, in which our laboratory has participated, utilized test samples of tumor tissue shipped at room temperature in ethanol, with excellent results on Southern blot. Unfortunately, however, ethanol is not the standard fixative used in surgical pathology—formalin is—and therefore it is unrealistic to expect that a significant proportion of specimens coming to the molecular pathology laboratory are going to be thus preserved; and that proportion will be even less for archival tissues.

Thus, the only practical approach at present is to encourage the surgical pathologists to set aside and freeze a portion of any specimen for which there is a possibility that Southern blot testing will be indicated. Since this is not an option for archival tissues and many specimens obtained postmortem, one will sometimes have to make do with formalin-fixed tissue and attempt to optimize the DNA extraction conditions for these suboptimal specimens. There are several such protocols in the literature, the details of which are beyond the scope of this chapter (Goelz et al. 1985; MacDonald and Cohen 1990; Scholefield et al. 1990; Dubeau et al. 1986). Even at their best, however, we and others (Wu et al. 1990) have obtained interpretable Southern blots from such specimens only about half the time.

As for other fixatives besides those already mentioned, there is universal agreement that tissues preserved in Bouin's fixative yield unsatisfactory results (Nuovo and Silverstein 1988). Results with mercuric chloride-containing fixatives such as Zenker's and B5 have also been disappointing (Mies et al. 1991), which is unfortunate since the latter fixative is commonly applied to lymphoma tumors because of its excellent morphologic preservation—and lymphomas are the tumor specimens most likely to require Southern blot analysis (see below). At the other end of the spectrum, cells fixed in Carnoy's solution (methanol–acetic acid), the standard fixative used for cytogenetic analysis, have yielded excellent Southern blot results (Ohyashiki et al. 1988), a propitious observation since specimens requiring karyotype study will often need molecular analysis as well.

2.3 Northern Blotting

There has been very little published experience with attempts to analyze RNA from fixed tissues by the extraction/electrophoresis/membrane transfer technique known as the Northern blot. In part this is due to the rather limited range of applications for Northern blot analyses in diagnostic molecular pathology at present. No doubt it also owes to the daunting challenge that would be inherent in any attempt to apply this procedure, a tricky technique even under the best conditions, to such suboptimal specimens. If DNA degradation is the primary concern in Southern blot analysis of embedded tissues, it stands to reason that RNA degradation in such specimens, owing to the ubiquity of endogenous and exogenous ribonucleases during processing, can only be worse. Indeed, Stanta and Scheider (1991) found average RNA fragment lengths of 100 to 200 nucleotrides in formalin-fixed, paraffin-embedded tissues. While this degree of degradation may still allow for specific mRNA detection by PCR or in situ hybridization, it is inadequate for readable Northern blots.

2.4 In Situ Hybridization

In situ hybridization—the application of DNA probes directly to fixed tissue sections on glass microscope slides—may be considered the paradigm for all the other techniques discussed in this chapter. For it alone was developed specifically for use on embedded specimens, and it is the technique which most literally epitomizes the fusion of molecular biology with classical morphologic pathology.

In situ hybridization involves no DNA extraction, amplification, electrophoresis, or blotting; in most instances it likewise involves no radioactive labeling. In this procedure, fixed tissues are left intact, save for the cutting of single 4–5 μ sections which are mounted on glass slides. The DNA probe and hybridization mix are added directly to the slide, essentially treating the glass as the equivalent of the nitrocellulose or nylon membrane in a Southern blot, and positive hybridization signals are observed under the microscope rather than by autoradiography. Cell and tissue morphology is thus preserved, and the molecular biological findings can be interpreted within the context of the original histopathologic architecture of the specimen. One can tell not only that a particular target nucleic acid sequence is present, but also in which particular cells it is contained or expressed.

The technique is highly favored by anatomic pathologists, since it requires no unusual processing of specimens and the interpretation of results is entirely histologic, not unlike immunohistochemistry or any other special stain. While the technique itself is quite tricky and requires molecular biology experience for probe preparation and troubleshooting, its interpretation demands little or no knowledge of nucleic acid biochemistry. Our laboratory was among the first to demonstrate its ready integration into routine surgical pathology practice (Grody et al. 1987).

The principal steps of in situ hybridization are listed below:

1. Deparaffinization of tissue sections
2. Protease digestion
3. Post-fixation in paraformaldehyde
4. Application of probe
5. Denaturation of DNA at high temperature (95–105°C)
6. Hybridization at physiologic temperature (37°C)
7. Post-hybridization stringency washes
8. Application of signal-generating system.

The major procedures are deparaffinization of tissue sections (typically with xylene), pretreatment with proteolytic enzymes, simultaneous denaturation of both probe and target DNA at high temperature, hybridization at physiological temperature, and application of one or another signal detection system. For the last of these, most pathology laboratories prefer a nonisotopic method. The most commonly used is biotin labeling of DNA probes by incorporation of biotinylated nucleotide derivatives (Langer et al. 1981). Hybridized probe can then be detected by complexing with avidin molecules linked to either a chromogenic enzyme, such as horseradish peroxidase or alkaline phosphatase, or fluorescent compounds. These will produce a brown, blue, or fluorescent color, respectively, over those cells containing the complementary target sequence of interest (Fig. 1). Alternatively, one can use radioactively labeled probes (^{35}S is preferred), which are then detected by overlaying the slide with a photographic emulsion and developing in the dark. In this case, positively hybridizing cells will be marked by the precipitation of black silver grains in the clear emulsion. Radioactive probes are said to be about tenfold more sensitive than biotin-labeled probes in this setting, though they are certainly more cumbersome to use. A more recent alternative has been offered by the introduction of chemiluminescent probe labels, which combine the convenience of biotin-labeled probes with a sensitivity virtually equivalent to that of radioactive probes. For ultrastructural localization of hybridized probe, the detection system can be conjugated with colloidal gold particles (Wolber et al. 1988). In situ hybridization can also be performed on cytology smear specimens (cells from body fluids applied directly to slides without embedding, usually followed by fixation in alcohol or formalin), with excellent results (Hilborne et al. 1987; Lian et al. 1991).

The most critical step in the in situ hybridization procedure, and the one which makes it so finicky, is the protease pretreatment. If this treatment is not long enough, the DNA probe will not be able to penetrate to its buried intracellular target, enmeshed in all the formalin-crosslinked proteins; and if it is too long, the DNA targets will lose their anchoring in the tissue and be washed off the slide (Grody et al. 1987). Estimating the most appropriate digestion time is difficult, since it will vary with each specimen, depending

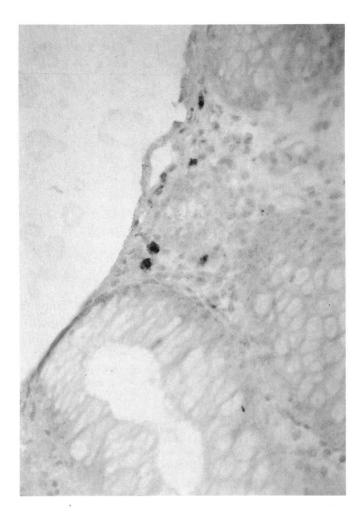

FIGURE 1. Example of in situ hybridization, using a biotin-labeled DNA probe for cytomegalovirus, applied to a formalin-fixed, paraffin-embedded section of a rectal biopsy from an AIDS patient. Mucosal cells containing CMV DNA sequences are darkly stained (× 335).

upon prior formalin fixation time and other factors. In our laboratory we routinely run each specimen in duplicate at two different protease times, in the hope that at least one will yield adequate results. We also include a post-fixation step in paraformaldehyde to re-anchor targets which have been loosened excessively back onto the slide.

As has been done for Southern blot analysis, various fixatives have been compared for their suitability for in situ hybridization. We and others (Nuovo and Richart 1989) have found buffered formalin to be ideal and have optimized our protocols for this routine fixative. As for Southern blots, Bouin's solution was found to be the least satisfactory.

In situ hybridization can be used for DNA targets (typically the DNA genomes of infecting viruses) or for RNA targets. The latter application is analogous to a Northern blot and is used either to detect the genomes of RNA viruses or to assess tissue-specific expression of human genes in differential diagnosis of poorly differentiated cancers (just as immunohistochemistry is used to detect their tissue-specific protein products). RNA appears to be fairly stable in embedded tissues once fixation has taken effect, though there can be substantial degradation prior to that time. In an ingenious variation on this theme, Tecott et al. (1988) have discovered that mRNA in fixed tissue sections can serve as a template for synthesis of a cDNA copy in situ, a process they dubbed "in situ transcription." To date this is the closest anyone has come to the elusive goal of combining in situ hybridization with some kind of amplification method (such as PCR) to increase its sensitivity.*

2.5 Polymerase Chain Reaction (PCR)

PCR is the most successfully applied molecular analytic technique for embedded samples, just as it is for most of the other sources of ancient DNA covered in this book, and for the same reasons. Since it works on even a single molecule of highly fragmented DNA, it is far less vulnerable to degradation-dependent failures and artifacts. Even if upwards of 99% of the sample is degraded down to single nucleotides, PCR can still act on the minority of target species remaining which are at least as long as the distance between the hybridization sites of the pair of oligonucleotide primers selected. The technique works well on DNA extracted from the whole embedded tissue or thick cuts (Imprain et al. 1987), from single 5-μ sections (Shibata et al. 1988a, b; Jackson et al. 1989), and from cytology, blood, or bone marrow smears and fixed chromosome spreads already on glass slides (Jackson et al. 1989; Jonveaux 1991; Hanson et al. 1990). Successful amplifications have been performed on specimens many decades old (Shibata et al. 1988b; Jackson et al. 1989). Our laboratory has demonstrated detection of rare retroviral sequences in postmortem tissues after only a very crude extraction from the paraffin blocks (Lai-Goldman et al. 1988).

With the introduction of an initial reverse transcriptase step, amplification of embedded RNA sequences, which are degraded too extensively to permit Northern blot analysis, can also be accomplished with great sensitivity, even from postmortem tissues years old which had not been fixed and embedded promptly after death (Redline et al. 1991; Weizsacker et al. 1991; Stanta and

---

Note added in proof. Methods for "in situ-PCR" have now been reported by a number of laboratories (Bagasra O, Seshamma T, Pomerantz RJ (1993) Polymerase chain reaction in situ: Intracellular amplification and detection of HIV-1 proviral DNA and other specific genes. J Immun Meth 158:131–145).

Scheider 1991). Amplified products of the embedded RNA or DNA templates are also amenable to hybridization with oligonucleotide probes (Imprain et al. 1987; Peiper et al. 1990) or to direct sequence analysis (Volkenandt et al. 1991).

As for the other techniques covered here, numerous authors have attempted to optimize conditions for PCR extraction and amplification, and have compared the suitability of various fixatives for the purpose. Treatments have ranged from simple deparaffinization in xylene followed by boiling (Shibata et al. 1988a; Coates et al. 1991), to extraction by sonication (Heller et al. 1991), to more involved extraction procedures involving multiple sequential washes followed by dialysis (Crisan et al. 1990). Comparison of fixatives has generally shown the most favorable results with ethanol, acetone, and formalin, and the least favorable with Zenker's, Bouin's, and B5 solutions, though the differences appear less compromising here than they do for Southern blots (Ben-Ezra et al. 1991; Rogers et al. 1990; Greer et al. 1991; Crisan et al. 1990).

2.6 Flow Cytometry

The ability to rapidly assess alterations in DNA content (ploidy) of large numbers of cells by the technique of flow cytometry has introduced a new prognostic tool in the pathologic study of tumors (Seckinger et al. 1989). In this procedure, ploidy and cell cycle analysis is performed by a fluorescence-activated cell sorter (FACS) which quantitates the cellular DNA index by measuring fluorescence resulting from the general DNA stain propidium (Vindelov et al. 1983). While this type of analysis was first worked out using single-cell suspensions derived from fresh tissues, embedded specimens have recently been shown to be an adequate source as well (Hedley et al. 1983). Single-cell suspensions are still required, but these can be prepared fairly easily by deparaffinization and rehydration, and in fact the procedure has recently been semiautomated (Cohen et al. 1991). As will now sound familiar, the most optimal fixatives for subsequent flow cytometric analysis are ethanol and formalin, while Zenker's-, Bouin's-, and B5-fixed tissues gave inferior results (Herbert et al. 1989; Alanen et al. 1989; Esteban et al. 1991).

3. Specific Diagnostic Applications

As embedded specimens have come to be increasingly amenable to molecular analysis, thanks to the imaginative adaptations of existing techniques for fresh tissue which have been reviewed in the previous section, the diagnostic applications for which they are used are also increasing—to the point that they now encompass virtually all of the major applications of molecular pathology. Clearly, a review of all relevant diagnostic applications in the broad field of molecular pathology is too expansive for this chapter, and so only a few prototypic examples in each disease category will be cited here. They are intended to be illustrative rather than comprehensive, a partial reflection of the wide range of diagnostic and research problems for which molecular analysis of embedded specimens—often the only kind of specimen available—can provide solutions.

3.1 Infectious Diseases

The first widespread application of molecular biology to diagnostic pathology was in the area of infectious diseases (Hilborne and Grody 1992), and the same is true for applications on embedded tissue. In fact, such applications are most necessary for fixed specimens, in which one no longer has the option of performing microbiological culture. In our institution, in situ hybridization for detection of viral infection in embedded biopsy tissues has become a routine diagnostic service. It is especially valuable in identifying infected cells which lack visible viral inclusions and thus would be overlooked by routine light microscopy (Myerson et al. 1984).

For diagnosis of infectious disease, one is most often looking for the presence or absence of a particular nucleic acid sequence foreign to human DNA, and so dot blots, PCR, and in situ hybridization are the most appropriate techniques. As already discussed, the latter has the advantage of localizing microbial sequences to particular cells within the tissue. This is often helpful in delineating the pathogenesis of the disease process and also in excluding the possibility that the positive signal is due to contamination with infected blood or exogenous material. For example, we were the first to demonstrate, using this method, the direct infection of cardiac myocytes by the human immunodeficiency virus (HIV) in the acquired immunodeficiency syndrome (AIDS) (Grody et al. 1990)—a finding which could not be rigorously proven merely by detecting HIV sequences in cardiac tissue by PCR or Southern blot or even by culture, as others had done previously (Calabrese et al. 1987; Flomenbaum et al. 1989). In a related study, we have used in situ hybridization to assess, in a series of archival tissues, the involvement of cytomegalovirus (CMV) in the pathogenesis of Kaposi's sarcoma (Grody et al. 1988).

On the other hand, the extreme sensitivity of PCR will often be necessary for detecting the presence of HIV sequences present in only 1 in 10,000 cells or less (Shibata et al. 1989). The ability to perform such studies on fixed material is also a plus for laboratory personnel safety when dealing with a deadly pathogen like HIV. And it allows for interesting retrospective studies on the prior prevalence and pathogenesis of unusual viruses such as HIV, human T-cell leukemia virus (Shibata et al. 1991), human papillomavirus (Shibata et al. 1988a; Liang et al. 1991), enteroviruses (Redline et al. 1991), and Epstein-Barr virus (Peiper et al. 1990). If PCR is the method chosen, detection of classical RNA viruses will require an initial reverse transcriptase step as mentioned earlier, whereas for detection of retroviruses one can often skip this step and directly target the proviral DNA (Shibata et al. 1991; Lai-Goldman et al. 1988).

Sometimes an embedded specimen is the only entree for making the correct diagnosis of viral infection. Some cases of hepatitis B virus (HBV) infection produce such low levels of the typical HBV serum antigens that they are not detectable by standard serologic studies. Such cases may be misdiagnosed as non-A, non-B hepatitis. Yet in situ hybridization will detect HBV DNA in the liver biopsies of such patients (Rijntjes et al. 1985). Similarly, PCR analysis of postmortem brain tissue has been used to confirm the diagnosis of herpesvirus encephalitis, even when collected at a nonculturable stage of the illness (Nicoll et al. 1991).

Despite the fact that Gram, acid-fast, and silver stains have been used on embedded tissues for many years for histochemical diagnosis of bacterial infection, molecular diagnostics for bacteria have been almost entirely restricted to fresh clinical samples suspended in buffer (solution hybridization using commercially available kits). In situ hybridization for infectious disease diagnosis has been limited mostly to viral detection, and for a long time it was assumed that the sturdy cell walls of bacteria would be impediments to probe hybridization. However, the technique has recently been used to identify *Helicobacter pylori* as a putative causative organism in biopsies of gastritis and gastric ulcer disease (Van den Berg et al. 1989). And our laboratory was the first to demonstrate successful targeting of bacterial ribosomal RNA in fixed tissue for in situ detection of *Legionella pneumophila* (Fain et al. 1991). Analogous techniques for protozoa and other unusual pathogens which are otherwise difficult to identify in fixed sections are being developed (Hayashi et al. 1990).

3.2 Neoplastic Diseases

In the field of cancer diagnosis, molecular biological techniques can be used to ascertain whether a tumor is monoclonal or polyclonal, whether it is benign or malignant, what its tissue of origin might be, and sometimes even its etiology. In surgical pathology it is not unusual to come across tumors that are so poorly differentiated that their tissue of origin cannot be readily deduced by morphologic examination alone. Often the dilemma does not become apparent until after the fact, when all excised tissue has been fixed in formalin. Analysis of gene expression patterns by mRNA in situ hybridization on the embedded specimen is one means to determine tumor type. For example, detection of mRNAs coding for specific hormones has been used in the differential diagnosis of endocrine tumors (Lloyd 1987).

In the area of lymphoid tumors, differential diagnosis between a benign reactive lesion and a malignant lymphoma often rests on determination of the lesion's clonality (or lack thereof). This is done by examining rearrangement patterns of the immunoglobulin and T-cell receptor genes on Southern blot. If the lesion is malignant, all the lymphocytes within it should be representative of a single clone and should exhibit an identical Southern blot pattern. Ideally we prefer to conduct such studies on fresh or frozen tissue, but when only fixed tissue is available, the methods discussed above for DNA extraction from embedded specimens can be attempted (Wu et al. 1990). Alternatively, amplification across these gene regions by PCR has been used to demonstrate clonality in such specimens (Wan et al. 1990; Goudie 1989).

Gene rearrangement on a more gross level—involving translocation of genetic material between different chromosomes—is also a prominent feature and useful marker of some malignancies. For example, a reciprocal translocation between chromosomes 9 and 22, which generates the "Philadelphia chromosome," is a consistent diagnostic and prognostic marker of chronic myelogenous leukemia. Traditionally detected by classical cytogenetics, it can be observed with greater sensitivity by Southern blotting with DNA probes derived from the breakpoint region on chromosome 22, and even more powerfully by PCR amplification across this region. The latter technique has been applied to detection of this and other cancer-specific translocations in both embedded tissue and glass slide smears, when such specimens were the only ones available (Hanson et al. 1990; Shibata et al. 1990a).

Additional tumor markers of diagnostic and prognostic value include the oncogenes and tumor suppressor genes. The former are associated with malignancy when they are either amplified at the DNA level or overexpressed at the RNA level, while one allele of the latter may be inactivated by mutation or completely deleted in certain tumors. These phenomena are best detected by Southern or Northern blot on fresh tissues, but they can also be performed on fixed tissue by Southern blot, in situ hybridization, or PCR (Cohen et al. 1988; Moerkerk et al. 1990; Lee et al. 1988; Bianchi et al. 1991). The utility of the last two techniques for such analysis is encouraging, should these assessments come to be accepted as part of the routine workup of common cancers, since they are far less cumbersome than Southern blots, especially for embedded tissue. Additional prognostic information can be derived from ploidy analysis by flow cytometry, a technique also applicable to fixed tissue, as described previously.

3.3 Genetic Diseases

While hereditary disease is arguably the most exciting domain at present for DNA diagnostics, it is the one least dependent on the use of embedded tissues as substrates. Many hereditary diseases are primarily biochemical rather than histopathologic in nature, so that surgical pathology services are not the site for their primary diagnosis in the first place. Most genetic screening is performed on living patients, many of whom (especially in the case of carrier screening) are themselves asymptomatic and therefore would have had no occasion for surgical removal of tissue. And the other major target for DNA-based genetic diagnosis—prenatal diagnosis of a fetus—is usually under such urgent time constraints (in order to allow for legal termination of pregnancy if desired) that archival studies of fixed tissue would be considered impractical and unnecessary, especially since fresh amniocentesis and chorionic villus specimens are readily available.

The one situation in which the issue of archival DNA study does come up is when diagnosis depends on restriction fragment-length polymorphism (RFLP) analysis in a family, and a family member essentially needed to complete the analysis is already deceased. RFLP studies are used to diagnose genetic diseases whose genes have not yet been identified. In the absence of a specific cDNA probe, we rely on the identification of an RFLP pattern that is so closely linked to the unknown disease gene that it virtually always co-segregates with the disease phenotype in the family being examined (Grody and Hilborne 1992). A minimum prerequisite for such a study is the presence in the family of at least one previously affected sibling to which the RFLP pattern of the fetus can be compared. If that individual is deceased, but had previously undergone either a biopsy or an autopsy, the only recourse may be to retrieve a paraffin block left over from that procedure and filed in the pathology department archives of the hospital at which it was performed. Such specimens are more likely to exist if the disorder in question is typically diagnosed by biopsy; for example, a paraffin-embedded muscle biopsy will often be available if a patient was diagnosed with muscular dystrophy (Grody et al. 1989b). In that case, any of the molecular procedures applicable to fixed tissue may be employed. If oligonucleotide primers are available to amplify across a key polymorphic restriction enzyme site, PCR may be the most efficacious method to use in this setting (Onadim and Cowell 1991). PCR may also be used for direct mutation detection in DNA from fixed tissue, but when such a specific test is available it is usually not necessary to examine material from deceased family members.

3.4 DNA Fingerprinting

This new technique, utilizing Southern blot analysis with hypervariable or minisatellite probes, or alternatively probes derived from the HLA-DQα locus, is now being applied to all manner of DNA samples in the forensic arena. Since such highly polymorphic genetic loci can now be accessed by PCR as well (Jeffreys et al. 1988; Walsh et al. 1991), DNA from even minute amounts of severely degraded forensic material can be used for individual identification. It goes without saying that the same technique can easily be applied to the relatively better preserved tissue in paraffin blocks (Bugawan et al. 1988). It has even been used for the purpose of identifying the blocks themselves, in cases where the specimens had been mislabeled (Shibata et al. 1990b).

4. Conclusion

This chapter has reviewed the use of the most common form of retained human specimens in clinical medicine—the paraffin-embedded tissue block—as a substrate for modern methods of molecular biological analysis for diagnostic purposes. The innovative techniques of nucleic acid extraction, hybridization, and amplification described herein have vastly expanded the array of clinical and biological information to be gleaned from these countless specimens on file in pathology departments all over the world. They have also caused us to alter our a priori view of the role of the old fixation and embedding process: from the destroyer of molecular biological information to its preserver. This realization, and continuing refinements in the technology, will continue to enhance the role and scope of molecular pathology well into the future— perhaps even to that time when the patient DNAs preserved in today's paraffin blocks may themselves be considered "ancient".

References

Alanen KA, Klemi PJ, Joensuu H, Kujari H, Pekkala E (1989) Comparison of fresh, ethanol-preserved, and paraffin-embedded samples in DNA flow cytometry. Cytometry 10:81–85

Ben-Ezra J, Johnson DA, Rossi J, Cook N, Wu A (1991) Effect of fixation on the amplification of nucleic acids from paraffin-embedded material by the polymerase chain reaction. J Histochem Cytochem 39:351–354

Bianchi AB, Navone NM, Conti CJ (1981) Detection of loss of heterozygosity in formalin-fixed paraffin-embedded tumor specimens by the polymerase chain reaction. Am J Pathol 138:279–284

Bramwell NH, Burns BF (1988) The effects of fixative type and fixation time on the quantity and quality of extractable DNA for hybridization studies. Exp Hematol 16:730–732

Bugawan TL, Saiki RK, Levenson CH, Watson RM, Erlich HA (1988) The use of non-radioactive oligonucleotide probes to analyze enzymatically amplified DNA for prenatal diagnosis and forensic HLA typing. Biotechnol 6:943–947

Calabrese LH, Proffitt MR, Yen-Lieberman B, Hobbs RE, Ratliff NB (1987) Congestive cardiomyopathy and illness related to the acquired immunodeficiency syndrome (AIDS) associated with isolation of retrovirus from myocardium. Ann Intern Med 107:691–692

Coates PJ, d'Ardenne AJ, Khan G, Kangro HO, Slavin G (1991) Simplified procedures for applying the polymerase chain reaction to routinely fixed paraffin wax sections. J Clin Pathol 44:115–118

Cohen PS, Seeger RC, Triche TJ, Israel MA (1988) Detection of N-*myc* gene expression in neuroblastoma tumors by *in situ* hybridization. Am J Pathol 131:391–397

Cohen C, Santoianni RA, Tickman RJ, Kennedy JC, DeRose PB (1991) Semiautomation of preparation of fixed paraffin-embedded tissue for DNA analysis. Anal Quant Cytol Histol 13:177–181

Conner BJ, Reyes AA, Morin C, Itakura K, Teplitz RL, Wallace RB (1983) Detection of sickle cell β-globin allele by hybridizaton with synthetic oligonucleotides. Proc Natl Acad Sci USA 80:278–282

Crisan D, Cadoff EM, Mattson JC, Hartle KA (1990) Polymerase chain reaction: amplification of DNA from fixed tissue. Clin Biochem 23:489–495 analysis of DNA extracted from formalin-fixed pathology specimens. Cancer Res 46:2964–2969

Esteban JM, Sheibani K, Owens M, Joyce J, Bailey A, Battifora H (1991) Effects of various fixatives and fixation conditions on DNA ploidy analysis: a need for strict internal DNA standards. Am J Clin Pathol 95:460–466

Fain JS, Bryan RN, Cheng L, Lewin KJ, Porter DD, Grody WW (1991) Rapid diagnosis of Legionella infection by a nonisotopic *in situ* hybridization method. Am J Clin Pathol 95:719–724

Feldman MY (1973) Reactions of nucleic acids and nucleoproteins with formaldehyde. Prog Nucl Acid Res Mol Biol 13:1–49

Flomenbaum M, Soeiro R, Udem SA, Kress Y, Factor SM (1989) Proliferative membranopathy and human immunodeficiency virus in AIDS hearts. J AIDS 2:129–135

Fraenkel-Conrat H (1954) Reactions of nucleic acid with formaldehyde. Biochim Biophys Acta 15:307–309

Goelz SE, Hamilton SR, Vogelstein B (1985) Purification of DNA from formaldehyde fixed and paraffin embedded human tissue. Biochem Biophys Res Commun 130:118–126

Goudie RB (1989) A strategy for demonstrating the clonal origin of small numbers of T lymphocytes in histopathological specimens. J Pathol 158:261–265

Greer CE, Peterson SL, Kiviat NB, Manos MM (1991) PCR amplification from paraffin-embedded tissues: effects of fixative and fixation time. Am J Clim Pathol 95:117–124

Grody WW, Hilborne LH (1992) Diagnostic applications of recombinant nucleic acid technology: genetic diseases. Lab Med 23:166–171

Grody WW, Cheng L, Lewin KJ (1987) *In situ* viral DNA hybridization in diagnostic surgical pathology. Hum Pathol 18:535–543

Grody WW, Lewin KJ, Naeim F (1988) Detection of cytomegalovirus DNA in classic and epidemic Kaposi's sarcoma by in situ hybridization. Hum Pathol 19:524–528

Grody WW, Gatti RA, Naeim F (1989a) Diagnostic molecular pathology. Mod Pathol 2:553–568

Grody WW, Hilborne LH, Spector EB (1989b) Clinical applications of molecular techniques in the diagnosis of inherited disease: the case of Duchenne muscular dystrophy. ASCP Clin Chem Check Sample 29(2):1–7

Grody WW, Cheng L, Lewis W (1990) Infection of the heart by the human immunodeficiency virus. Am J Cardiol 66:203–206

Hanson CA, Holbrook EA, Sheldon S, Schnitzer B, Roth MS (1990) Detection of Philadelphia chromosome-positive cells from glass slide smears using the polymerase chain reaction. Am J Pathol 137:1–6

Haselkorn R, Doty P (1960) The reaction of formaldehyde with polynucleotides. J Biol Chem 236:2738–2745

Hayashi Y, Watanabe J-I, Nakata K, Fukayama M, Ikeda H (1990) A novel diagnostic method of *Pneumocystis carinii: in situ* hybridization of ribosomal ribonucleic acid with biotinylated oligonucleotide probes. Lab Invest 63:576–580

Hedley DW, Friedlander ML, Taylor IW, Rugg CA, Musgrove EA (1983) Method for analysis of cellular DNA content of paraffin-embedded pathological material using flow cytometry. J Histochem Cytochem 31:1333–1335

Heller MJ, Burgart LJ, TenEyck CJ, Anderson MEk, Greiner TC, Robinson RA (1991) An efficient method for the extraction of DNA from formalin-fixed, paraffin-embedded tissue by sonication. BioTechniques 11:372–377

Herbert DJ, Nishiyama RH, Bagwell CB, Munson ME, Hitchcox SA, Lovett EJ (1989) Effects of several commonly used fixatives on DNA and total nuclear protein analysis by flow cytometry. Am J Clin Pathol 91:535–541

Hilborne LH, Grody WW (1992) Diagnostic applications of recombinant nucleic acid technology: infectious diseases. Lab Med 23:89–94

Hilborne LH, Nieberg RK, Cheng L, Lewin KJ (1987) Direct in situ hybridization for rapid detection of cytomegalovirus in bronchoalveolar lavage. Am J Clin Pathol 87:766–769

Hsu H-C, Pen S-Y, Shun C-T (1991) High quality of DNA retrieved for Southern blot hybridization from microwave-fixed, paraffin-embedded liver tissues. J Virol Meth 31:251–262

Impraim CC, Saiki RK, Erlich HA, Teplitz RL (1987) Analysis of DNA extracted from formalin-fixed, paraffin-embedded tissues by enzymatic amplification and hybridization with sequence-specific oligonucleotides. Biochem Biophys Res Commun 142:710–716

Jackson DP, Bell S, Payne J, Lewis FA, Sutton J, Taylor GR, Quirke P (1989) Extraction and amplification of DNA from archival haematoxylin and eosin sections and cervical cytology Papanicalaou smears. Nucl Acids Res 17:10134

Jackson V (1978) Studies on histone organization in the nucleosome using formaldehyde as a reversible cross-linking agent. Cell 15:945–954

Jackson V, Chalkey R (1974) Separation of newly synthesized nucleohistone by equilibrium centrifugation in cesium chloride. Biochemistry 13:3952–3956

Jeffreys AG, Wilson V, Neumann R, Keyte J (1988) Amplification of human minisatellites by the polymerase chain reaction: towards DNA fingerprinting of single cells. Nucl Acids Res 16:10953–10971

Jonveaux P (1991) PCR amplification of specific DNA sequences from routinely fixed chromosomal spreads. Nucl Acids Res 19:1946

Junqueira LC, Carneiro J (1971) Basic Histology, 3rd ed. Los Altos, Cal.: Lange Medical Publications, pp. 2–3

Lai-Goldman M, Lai E, Grody WW (1988) Detection of human immunodeficiency virus infection in formalin-fixed, paraffin-embedded tissues by DNA amplification. Nucl Acids Res 16:8191

Langer PR, Waldrop AA, Ward DC (1981) Enzymatic synthesis of biotin-labeled polynucleotides: novel nucleic acid affinity probes. Proc Natl Acad Sci USA 78:6633–6637

Lee JH, Lee DH, Park SS, Seok SE, Lee JD (1988) Oncogene expression detected by in situ hybridization in human primary lung cancer. Chest 94:1046–1049

Liang X-M, Wieczorek RL, Koss LG (1991) In situ hybridization with human papillomavirus using biotinylated DNA probes on archival cervical smears. J Histochem Cytochem 39:771–775

Lloyd RV (1987) Use of molecular probes in the study of endocrine diseases. Hum Pathol 18:1199–1211

MacDonald MR, Cohen BB (1990) Southern blot analysis of DNA extracted from paraffin wax-embedded tissue. Dis Markers 8:341–345

Mark A, Trowell H, Dyall-Smith ML, Dyall-Smith DJ (1987) Extraction of DNA from formalin-fixed paraffin-embedded pathology specimens and its use in hybridization (histoblot) assays. Application to the detection of human papillomavirus DNA. Nucl Acids Res 15:8565

McFadden GI, Bonig I, Cornish EC, Clarke AE (1988) A simple fixation and embedding method for use in hybridization histochemistry on plant tissues. Histochem J 20:575–586

McGhee JD, Von Hippel PH (1977) Formaldehyde as a probe of DNA structure. 4. Mechanism of the initial reaction of formaldehyde with DNA. Biochemistry 16:3276–3293

Mies C, Houldsworth J, Chaganti RSK (1991) Extraction of DNA from paraffin blocks for Southern analysis. Am J Surg Pathol 15:169–174

Moerkerk PT, Kessels HJ, ten Kate J, de Goeij FPM, Bosman FT (1990) Southern and dot blot analysis of DNA from formalin-fixed, paraffin-embedded tissue samples from colonic carcinomas. Virch Arch B Cell Pathol 58:351–355

Myerson D, Hackman RC, Nelson JA, Ward DC, McDougall JK (1984) Widespread presence of histologically occult cytomegalovirus. Hum Pathol 87:766–769

Nicoll JAR, Maitland NJ, Love S (1991) Use of the polymerase chain reaction to detect herpes simplex virus DNA in paraffin sections of human brain at necropsy. J Neurol Neurosurg Psychiat 54:167–168

Nuovo GJ, Richart RM (1989) Buffered formalin is the superior fixative for the detection of HPV DNA by *in situ* hybridization analysis. Am J Pathol 134:837–842

Nuovo GJ, Silverstein SJ (1988) Comparison of formalin, buffered formalin, and Bouin's fixation on the detection of human papillomavirus deoxyribonucleic acid from genital lesions. Lab Invest 59:720–728

Ohyashiki JH, Ohyashiki K, Toyama K (1988) Analysis of DNA from cells fixed in Carnoy's solution for cytogenetic study. Cancer Genet Cytogenet 34:159–163

Onadim Z, Cowell JK (1991) Application of PCR amplification of DNA from paraffin embedded tissue sections to linkage analysis in familial retinoblastoma. J Med Genet 28:312–316

Peiper SC, Myers JL, Broussard EE, Sixbey JW (1990) Detection of Epstein-Barr virus genomes in archival tissues by polymerase chain reaction. Arch Pathol Lab Med 114:711–714

Radosevich JA, Maminta LD, Rosen ST, Gould VE (1991) The amount of paraffin-embedded tissue needed for DNA molecular analysis: a rapid extraction procedure. Lab Med 22:543–546

Redline RW, Genest DR, Tycko B (1991) Detection of enteroviral infection in paraffin-embedded tissue by the RNA polymerase chain reaction technique. Am J Clin Pathol 96:568–571

Rijntjes PJM, Van Ditzhuijes TJM, Van Loon AM, Van Haelst UJGM, Bronkhorst FB, Yap SH (1985) Hepatitis B virus DNA detected in formalin-fixed liver specimens and its relation to serologic markers and histopathologic features in chronic liver disease. Am J Pathol 120:411–418

Rogers BB, Alpert LC, Hine EAS, Buffone GJ (1990) Analysis of DNA in fresh and fixed tissue by the polymerase chain reaction. Am J Pathol 136:541–548

Scholefield JH, McIntyre P, Palmer JG, Coates PJ, Shepherd NA, Northover JMA (1990) DNA hybridisation of routinely processed tissue for detecting HPV DNA in anal squamous cell carcinomas over 40 years. J Clin Pathol 43:133–136

Seckinger D, Sugarbaker E, Frankfurt O (1989) DNA content in human cancer: Application in pathology and clinical medicine. Arch Pathol Lab Med 113:619–626

Seto E, Yen TSB (1987) Detection of cytomegalovirus infection by means of DNA isolated from paraffin-embedded tissues and dot hybridization. Am J Pathol 127:409–413

Shibata D, Arnheim N, Martin WJ (1988a) Detection of human papillomavirus in paraffin-embedded tissue using the polymerase chain reaction. J Exp Med 167:225–230

Shibata D, Martin WJ, Arnheim N (1988b) Analysis of DNA sequences in forty-year-old paraffin-embedded thin-tissue sections: a bridge between molecular biology and classical histology. Cancer Res 48:4564–4566

Shibata D, Brynes RK, Nathwani B, Kwok S, Sninsky J, Arnheim N (1989) Human immunodeficiency viral DNA is readily found in lymph node biopsies from sero-positive individuals: analysis of fixed tissue using the polymerase chain reaction. Am J Pathol 135:697–702

Shibata D, Hu E, Weiss LM, Brynes RK, Nathwani BN (1990a) Detection of specific t(14;18) chromosomal translocations in fixed tissues. Hum Pathol 21:199–203

Shibata D, Namiki T, Higuchi R (1990b) Identification of a mislabeled fixed specimen by DNA analysis. Am J Surg Pathol 14:1076–1078

Shibata D, Tokunaga M, Sasaki N, Nanba K (1991) Detection of human T-cell leukemia virus type I proviral sequences from fixed tissues of seropositive patients. Am J Clin Pathol 95:536–539

Smith LJ, Braylan RC, Nutkis JE, Edmundson KB, Downing JR, Wakeland EK (1987) Extraction of cellular DNA from human cells and tissues fixed in ethanol. Anal Biochem 160:135–138

Stanta G, Schneider C (1991) RNA extracted from paraffin-embedded human tissues is amenable to analysis by PCR amplification. BioTechniques 11:304–308

Tecott LH, Barchas JD, Eberwine JH (1988) In situ transcription: specific synthesis of complementary DNA in fixed tissue sections. Science 240:1661–1664

Tokuda Y, Nakamura T, Satonaka K, Maeda S, Doi K, Baba S, Sugiyama T (1990) Fundamental study on the mechanism of DNA degradation in tissues fixed in formaldehyde. J Clin Pathol 43:748–751

Van den Berg FM, Zijlmans H, Langenberg W, Rauws E, Schipper M (1989) Detection of *Campylobacter pylori* in stomach tissue by DNA in situ hybridization. J Clin Pathol 42:995–1000

Vindelov LL, Christensen IJ, Nissen NI (1983) A detergent-trypsin method for the preparation of nuclei for flow cytometric DNA analysis. Cytometry 3:323–327

Volkenandt M, McNutt NS, Albino AP (1991) Sequence analysis of DNA from formalin-fixed, paraffin-embedded human malignant melanoma. J Cutan Pathol 18:210–214

Walsh PS, Metzger DA, Higuchi R (1991) Chelex 100 as a medium for simple extraction of DNA for PCR-based typing from forensic material. BioTechniques 10:506–513

Wan JH, Trainor KJ, Brisco MJ, Morley AA (1990) Monoclonality in B cell lymphoma detected in paraffin wax embedded sections using the polymerase chain reaction. J Clin Pathol 43:888–890

Warford A, Pringle JH, Hay J, Henderson SD, Lauder I (1988) Southern blot analysis of DNA extracted from formol-saline fixed and paraffin wax embedded tissue. J Pathol 154:313–320

Weizsacker FV, Labeit S, Koch HK, Oehlert W, Gerok W, Blum He (1991) A simple and rapid method for the detection of RNA in formalin-fixed, paraffin-embedded tissues by PCR amplification. Biochem Biophys Res Commun 174:176–180

Wolber RA, Beals TG, Lloyd RV, Maasab HF (1988) Ultrastructural localization of viral nucleic acid by in situ hybridization. Lab Invest 59:144–151

Wu AM, Ben-Ezra J, Winberg C, Colombero AM, Rappaport H (1990) Analysis of antigen receptor gene rearrangements in ethanol and formaldehyde-fixed, paraffin-embedded specimens. Lab Invest 63:107–114

Fixed and Embedded Samples
6 DNA from Amber Inclusions

GEORGE O. POINAR JR., HENDRIK N. POINAR, and
RAUL J. CANO

1. Introduction

Amber, together with jet and coral, is one of the few "organic gems"; its use as an ornament and as an art medium throughout history has been the subject of several books (Hunger 1979; Rice 1980; Fraquet 1987). Amber is fossilized tree resin which can vary in age from 4 to 225 million years. The exact process by which plant resins become amber is unknown. However, it involves evaporation of essential oils and polymerization, which result in the hardening of the resin. Most amber has a hardness between 2.0 and 2.5 on the Mohl's scale, a refractive index of 1.5–1.6, a specific gravity between 1.06 and 1.10, and a melting point of 250–300°C. Of course, a very important prerequisite for fossilization is that the resin be resistant to microbial decomposition, since it will be exposed to soil, rock, and usually seawater for millions of years. Few plant genera are known to produce resin that can withstand the physical and biological forces on the earth's surface for such a long time (Poinar 1992).

Today, most commercially available amber comes from either the Baltic sea region or from the Dominican Republic. Many of the pieces available to the buyer may have insect or plant remains within them; such pieces have been the object of marvel and scientific examination over the centuries (Table 1). An amazing diversity of life forms have been studied and described from amber deposits around the world (Poinar 1992).

The recent extraction and characterization of DNA from extinct Dominican (Cano et al. 1992a,b; DeSalle et al. 1992) and Lebanese amber insects (Cano et al. 1993) and plants (Poinar et al., 1993), shown in Table 2, opens up the diversity of life in amber to DNA analysis and studies on molecular evolutionary change through time. There are several highly fossiliferous amber deposits around the world, varying in age from 20 to 130 million years (Table 1). Dominican Republic amber is one of these, and is presently known to contain representatives of seven classes of arthropods. The largest class is the Insecta, with representatives of 23 orders and 229 families fossilized in amber (Poinar 1992). In addition to these arthropods there are

TABLE 1. Significant amber deposits containing fossil inclusions

Amber deposits	Location	Approximate age (millions of years)	Types of inclusions
Dominican	Dominican Republic	25–40	Invertebrate, Vertebrate and Plant
Baltic	Northern Europe	40	Invertebrate, Vertebrate and Plant
Mexican	Chiapas, Mexico	22–26	Invertebrate, Vertebrate and Plant
Canadian	Manitoba, Alberta	70–80	Invertebrate, Vertebrate and Plant
Taimyr	Russian Arctic	78–115	Invertebrate, Vertebrate and Plant
Middle East (Lebanese)	Lebanon, Israel, Jordan	120–135	Invertebrate and Plant

For further details about the characteristics of and inclusions in these amber deposits, see Poinar (1992).

members of several other invertebrate phyla (Mollusca; Annelida; Rotifera; Nematoda) as well as the remains of four classes of vertebrates and a range of plants.

2. Historical Aspects

Although extracting DNA from amber inclusions is a relatively recent endeavor which makes crucial use of the rapid developments in the area of cell and molecular biology (especially PCR), reviving simple life forms from fossilized resin has been a dream for some time.

In 1920, Galippe made an attempt to culture bacteria from Baltic amber (Galippe 1920). He surface-sterilized pieces of amber, ground them up, and then placed them in culture media. He recovered several bacterial species, but they all belonged to ubiquitous groups that one would expect to encounter as contaminants. There was no way to prove that these cells had been embedded in the amber and had undergone suspended animation for 40 million years.

The first attempt to extract DNA from amber insects was made in 1983 when A. Wilson, R. Higuchi, G. Poinar, Jr., and R. Hess collaborated on a project which involved scraping the body contents of eight Dominican amber insects into buffer and extracting the DNA. In two of the eight insects, extracts made from the remains contained templates capable of directing the synthesis of radioactive complementary DNA in the presence of exogenous primers, purified DNA polymerase 1 from *E. coli* and ^{32}P-labeled nucleoside triphosphates. Only brief mention was made of these attempts (Higuchi and Wilson 1984), and no hybridization experiments were done to determine whether the results were due to human contamination. The first fully reported

TABLE 2. Biological inclusions in amber investigated for DNA

Type of material	Identification	Number of specimens	Amber source	Extraction	DNA stages amplification	Sequencing	Reference
Insect	*Proplebeia dominicana* (Apidae: Hymenoptera) Fig. 2a, b	5	Dominican amber	+	+	+	Cano et al. 1992a, b
Insect	*Orthellia* spp. (Mycetophilidae: Diptera) (Fig. 2b)	4	Dominican amber	+	+	+	Poinar, H.N. unpublished data
Insect	Curculionoidea (Coleoptera)	1	Lebanese amber	+	+	+	Cano et al., 1993
Insect	*Mastotermes electrodominicus* (Mastotermitidae: Isoptera)	1	Dominican amber	+	+	+	DeSalle et al., 1992
Plant	*Hymenaea protera* (Leguminoseae) (Fig. 2c)	1	Dominican amber	+	+	+	Poinar, et al., 1993
Vertebrate	*Anolis* spp. (eguanidae: Reptilia)	1	Dominican amber	+	+	+	Poinar, H.N. in preparation

extraction and partial characterization of DNA from amber inclusions was from stingless bees in Dominican amber (Cano et al. 1991a). Hybridization studies showed that the extracted DNA was not from bacterial or human contamination. A second report provided information on the amplification and sequencing of this bee DNA, which confirmed the hymenopteran origin of the extracts (Cano et al. 1991b). Further studies with other insects and plants are listed in Table 2.

3. Preservative Qualities of Amber

Although various scientists had noticed the presence of dried tissue in arthropods embedded in amber, the degree of preservation of amber-entombed tissue and cells was first demonstrated by Poinar and Hess (1982, 1985) when they examined a Baltic amber fungus gnat (Mycetophilidae: Diptera) under the electron microscope (Fig. 1). A strip of hypodermal and muscle tissue adjacent to the cuticle of the gnat was exceptionally well preserved. Distinct epidermal cell nuclei were observed, with areas of electron-dense nucleoplasm probably representing chromatin as well as myelin swirls and smooth endoplasmic reticulum. In addition, autophagic vacuoles, lysosomes, mitochondria, lipid droplets, and ribosomes could be identified in the epidermal cells. The muscle cells contained distinct bundles of muscle fibrils and long mitochondria, with prominent cristae interspersed between the myofibrils. Tracheae and tracheoles were closely associated with the muscle surface or within the cell surface, and still contained an exterior plasma membrane and an inner lipoprotein cuticulin layer.

An analysis of the Baltic amber gnat tissue suggested that two major factors were responsible for the high degree of preservation. The first was a physical phenomenon known as "inert dehydration": components of the resin withdraw the moisture from the original tissue, resulting in an extreme degree of mummification. The second factor was the antimicrobial action imparted by the resin which, together with the absence of oxygen needed by aerobic organisms, protected the tissues from decay. Since most amber pieces have been deposited in layers of sandstone or limestone for the longest part of their existence, the tissues of entombed organisms essentially have been stored in an oxygen-depleted environment. The ability to retrieve DNA from insects locked into resin up to 120–135 million years ago (Table 2) attests to the remarkable preservative properties of amber.

4. Locating and Selecting Specimens for DNA Extraction

Amber with inclusions can be obtained from state and national museums or private collections, or purchased directly from stores or amber dealers. It is important to obtain samples from dependable dealers that can provide

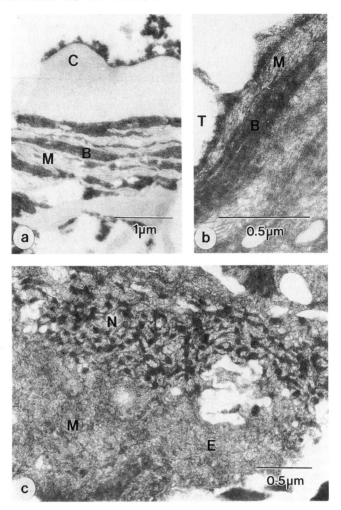

FIGURE 1a–b. Electron micrographs of tissue from an amber embedded fossil fungus gnat. (*a*) In this micrograph the cuticle (C) is present and beneath are layers of muscle bundles (B) and mitochondria (M). (*b*) Included here are the edge of a trachea (T) with muscle bundles (B) and mitochondria (M). (*c*) The nucleus (N) is prominent in this micrograph with condensed chromatin. In the cytoplasm are membranous profiles, probably representing endoplasmic reticulum (E) and small mitochondria (M). Bars represent 1 μm. Photos courtesy of Roberta Hess-Poinar.

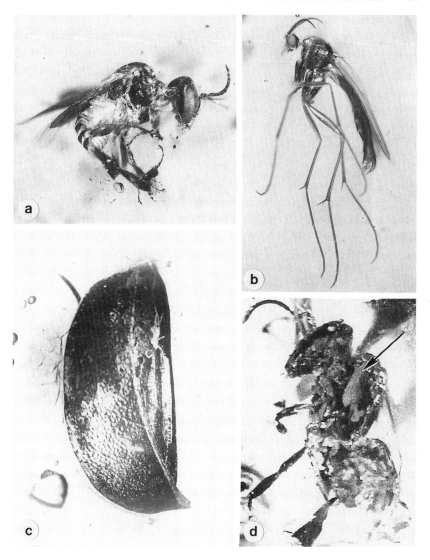

FIGURE 2a–d. Representative organisms in Dominican amber from which DNA has been extracted. (a) The stingless bee, *Proplebeia dominicana* (Apidae: Hymenoptera). (b) A fungus gnat (Mycetophilidae: Diptera). (c) Leaf of the extinct tree, *Hymenaea protera* (Leguminoseae). (d) A piece of Dominican amber that has been cracked open in order to remove tissue from the body of a fossilized stingless bee (*Proplebeia dominicana*) according to the procedure outlined in Cano et al. (1992 a, b) and the present work. Arrow indicates preserved muscle bundles in the thorax of the fossil bee.

reliable data on the source and sometimes even the specific mine where the amber originated. Purchasing amber in shops can be risky because the clerk may not be certain about the origin of the samples and there is always the chance of obtaining a fake. Two types of fakes are commonly found on the market today: the first is represented by a range of plastics and the second consists of natural resins of a relatively young age (copal). Fakes can be identified by relatively simple physical and chemical tests (Poinar 1982), or with sophisticated instrumentation that produce infrared or nuclear magnetic spectra that can be compared with previously run spectra from authentic samples. Fakes have modern life forms embedded in them, whereas the great majority of the life forms found in amber from the various sources listed in Table 1 are now extinct at the species or a higher level.

The amount of preserved tissue in amber insects varies greatly, even for individuals of the same species, depending on the conditions under which the specimens were preserved. The best samples are those in which the insects appear to have been completely and immediately covered by the sap, thereby allowing components of the resin to act on the fresh tissues. Frequently, however, insects became stuck and died on the surface of the resin, thus undergoing decomposition and autolysis before being covered by a subsequent resin flow. Under these conditions, much of the original tissue will be gone and any remaining DNA greatly damaged. Normally, but not always, a specimen that is perfectly preserved on the outside will contain some tissue remains.

In thin-walled, soft-bodied insects such as bark lice (Psocoptera), some bugs (Homoptera), and many flies (Diptera), microscopic examination of the specimen will reveal the presence of tissue in the abdomen, thorax, or legs. It may be necessary to re-polish the piece in order to better view the specimen, but care must be taken not to cut any of the appendages (legs, antennae), which could potentially allow air and moisture or contaminants to enter the tissues. For DNA studies, it is important to keep the piece as clean as possible.

After an inclusion is found which contains tissue remains and is selected for DNA extraction, the following procedure can be followed.

5. Experimental Procedures

5.1 Sterilization

Considering the remarkable ability of PCR to amplify even the smallest amount of template, it is crucial that the amber samples are properly sterilized before they are opened.

The amber pieces chosen for DNA extraction are placed in 2% phosphate-buffered (pH 7.0) glutaraldehyde for 24 hours. Immediately prior to the DNA extraction process, the amber pieces are rinsed thoroughly in sterile distilled water and placed into 5% sodium hypochlorite to remove any exogenous nucleic acid contamination. The pieces are rinsed again in sterile distilled water, immersed in 70% ethanol, and briefly flamed.

It is essential that a suitable process for validating the sterilization and decontamination of the amber be established. Aliquots of the sterile water rinses as well as sterile water eluates of amber chips (after the pieces are cracked) from internal and external areas of the amber should be used as templates in PCR amplification studies using the appropriate primers. Other validation steps may be necessary depending upon the goals of the study.

5.2 Extraction

Two methods for removing tissue from amber inclusions are feasible. In the first method, as described by Cano et al., the sterilized piece of amber is placed into a sterile petri dish on a dissecting scope in a class II laminar flow hood. Liquid nitrogen is poured on the sample and when it has evaporated, a few drops of hot, sterile physiological saline are poured over the piece. This temperature transition usually fractures the piece along the plane of the inclusion where it has been weakened, and enables it to be pried open. It is important to choose pieces with fracture lines running parallel to the body of the inclusion, to permit easy extraction of the tissue once it is exposed. The amber piece is transferred to another sterile petri dish and carefully opened with 27-gauge needles. When the tissue is exposed, the amber piece is glued to the petri dish to facilitate removal of the tissue.

Once the amber is securely fastened, a second set of 27-gauge needles are dipped into the chelex solution and used to remove the dried tissue. Dry needles are ineffective in removing tissue due to the static charge which causes the tissue to separate over the petri dish. For insects, tissue from the head, thorax, or abdomen is removed and placed in 500 μl of 5% chelex (Bio Rad, Richmond CA) or a solution of guanidium isothiocyanate and glass milk (Bio 101, La Jolla, CA) which has been shown to be successful in the extraction of ancient DNA (Cano and Poinar, 1993). Amber fragments can be placed in broth (BBL, Cockeyesville, MD) in an attempt to recover bacterial endospores still entombed in the amber, and for later hybridization studies. Since the abdomen may contain partially digested materials in the gut, there may be a greater chance of contamination in some insects. Cultures of amber pieces devoid of tissue should be made to evaluate surface sterility.

Tissues from various body locations (head, thorax, abdomen) are placed into two separately labeled 1.5-ml eppendorfs with the chelex solution. The tubes are rocked in a 60°C incubator overnight or until the tissue appears to have dissolved. The mixture is then vortexed for 5 sec and placed into a dry bath (95–100°C) for 10 min. The tubes are vortexed again for 5 sec and centrifuged briefly. Two sets of serial dilutions are made as follows. First, 20 μl of chelex supernatant is mixed with 70 μl of deionized water; from this solution 10 μl is removed and mixed with 90 μl of deionized water. Ten microliters of the latter is removed and used as a template for PCR.

We have developed a less destructive method for extracting tissue. Using a micro drill with bit sizes under 1mm diameter, a small hole is drilled into the amber adjacent to the inclusion. A 23 gauge needle is then used to pierce through the remaining thin layer of amber into the inclusion. The dried tissue is removed with a 27 gauge needle attached to a vacuum pump with an adjustable valve controlling suction through the needle. Tissue siphoned through the needle is deposited in a microcentrifuge tube containing 200–500 μl of 5% chelex.

The latter method has two advantages over the previously-described protocol. First, it does not allow the tissue to be exposed to the surrounding environment, lowering the probability that environmental DNA will contaminate the sample. Second, the small hole in the amber may be resealed with synthetic resin so that the rest of the organism is essentially untouched and the aesthetic qualities of the amber piece and specimen are preserved.

5.3 DNA Hybridization

Hybridization assays can be used to ensure that the DNA removed from the samples is actually DNA from the inclusion rather than from external or internal contamination.

5.3.1 Probe Preparation and Labeling

Genomic DNA extracted from tissues can be used as probe in reverse sample genome probing (Cano et al. 1992a; Voordouw et al. 1991) in order to quantify the degree of DNA sequence homology between target and probe DNA. In this assay, each DNA sample is used both as probe and as target. A control panel of possible contaminating DNAs is established for the evaluation of the DNA sample removed from amber.

5.4 DNA Amplification

Amplification of aDNA has received much attention recently, especially in regards to inhibitors such as hematin and porphyrins. Bovine serum albumin (BSA) has proven very successful in blocking inhibition due to its strong affinity for porphyrins. Adding albumin to the extraction portion of the chelex reaction has proven inefficient since it appears that the chelex absorbs the albumin. However, the addition of BSA (4 μl/ml) to the master PCR mix has been shown to be efficient in amplifying aDNA (Pääbo 1990). It has also been successful in the amplification of modern plant tissue known to contain porphyrins which inhibit the PCR.

We are using a modified "hot start" PCR method which yields large quantities of amplified product by preventing random annealing of primers to template DNA. This method also reduces the inaccuracies arising from delivering < 1 μl of viscous *Taq* polymerase in storage buffer. All dilutions and reagent mixes are conducted in tubes or microwell plates in an ice bath, or in a pre-chilled thermal cycler. Once the solutions are made the samples are placed in a freezer while the heating block of the thermal cycler warms to 80°C. At that time the samples are placed back in the thermal cycler for 3 min. This step is followed by the appropriate temperature cycling protocol. Reagent concentration and temperatures (particularly annealing temperatures) must be optimized to obtain maximum yields.

Primers selected from the 18S rRNA gene of *Saccharomyces cerevisiae*, *Dictyostelium discoideum*, and *Stylonichia pustulata* have included NS1, NS2, NS3, NS4, and NS19. All have proven extremely successful in amplifying ancient remains, due to their highly conserved nature. We have designed specific ribosomal primers to avoid amplification of contaminating DNA.

5.5 Sequencing of Amplification Products

Several methods for determining the deoxynucleotide sequence of specific PCR products have been published (Ellingboe and Gyllensten 1992). Cano et al. (1992b) sequenced purified double-stranded PCR products by enzymatic amplification using a fmol™ DNA sequencing kit (Promega, Madison, WI) in the presence of α-thio[^{35}S]dATP. Electrophoresis of sequencing products was performed in a 6% sequencing gel consisting of 5.7% (w/v) acrylamide, 0.3% (w/v) N,N'-methylenebisacrylamide, 7 M urea, 100 mM Tris-borate (pH 8.3), and 1 mM EDTA using a model S2 sequencing gel electrophoresis apparatus (Gibco BRL, Gaithersburg, MD). Autoradiographs of air-dried sequencing gels were made using XAR X-ray film (Kodak, Rochester, MN). Sequences were inferred from autoradiographs using a gel reader (IBI/Kodak, New Haven, CT) and then aligned with MacVector 3.5 Sequence Analysis software (IBI/Kodak, New Haven, CT).

6. Prospects

Future studies on the characterization of DNA from amber inclusions are very promising. Not only is amber available for a wide age range but also featuring a wide variety of invertebrate, plant, and vertebrate inclusions from which to select (although vertebrate fossils in amber often are priced in the tens of thousands of dollars and are difficult to acquire).

Sequences from amber inclusions allow us to include extinct organisms that existed millions of years ago in molecular phylogeny studies. These results enable us to essentially look at genes which have stopped evolving 20, 40, and 120 million years ago, and answer fundamental questions about the substitution rate for specific genes. They may also shed light on biogeographic questions raised about certain endemic or dispersed species. How much of the original genome still exists of various life forms in amber is not known. Certain environmentally resistant stages, such as spores, might well have the bulk of their DNA preserved whereas DNA found in less durable tissues such as those of insects and plants may have suffered much more damage. Studies with DNA from organisms preserved in amber will be greatly enhanced when damage (inflicted on the DNA during the preservation process) can be repaired.

The recent studies described here on the recovery of DNA from amber inclusions significantly extend the limits of DNA recovery beyond previous reports and demonstrate the almost unbelievable shelf-life of DNA. What most scientists thought impossible some ten years ago, when DNA extraction studies on amber inclusions were first attempted, has now become a reality.

References

Cano RJ, Nelson ML (1992) Fluorescent DNA hybridization assay increases sensitivity of clinical pathogen detection. Millipore Bioforum 2:1–7

Cano RJ, Poinar H, Poinar Jr GO (1992a) Isolation and partial characterization of DNA from the bee *Proplebeia dominicana* (Apidae: Hymenoptera) in 25–40 million year old amber. Med Sci Res 20:249–251

Cano RJ, Poinar HN, Roubik D, Poinar Jr GO (1992b) Enzymatic amplification and nucleotide sequencing of portions of the 18s rRNA gene of the bee *Proplebeia dominicana* (Apidae: Hymenoptera) isolated from 25–40 million year old Dominican amber. Med Sci Res 20:619–623

Cano RJ, Torres J, Klem R. Palomares JC (1992c) DNA hybridization assay using AttoPhos™, a fluorescent substrate for alkaline phosphatase. Bio Techniques 12:264–269

Cano RJ, Poinar HN, Pieniazek NJ, Acra A, Poinar Jr GO (1993) Amplification and sequencing of DNA from a 120–135 million year old weevil. Nature 363:536–538.

Cano RJ, Poinar HN (1993) A rapid and simple method for extracting DNA from fossil and museum specimens suitable for the Polymerase Chain Reaction. Biotechniques (in press)

DeSalle R, Gatesy J, Wheeler W, Grimaldi D (1992) DNA sequences from a fossil termite in Oligo-Miocene amber and phylogenetic implications. Science 257:1933 1936

Ellingboe J, Gyllensten UB (1992) *The P.C.R. Technique: DNA Sequencing*. Natick, Mass.: Eaton

Fraquet H (1987) *Amber*. London: Butterworths

Galippe V (1920) Recherches sur la résistance des microzymas à l'action du temps et sur leur survivance dans l'ambre. Comp Rendu Acad Sci (Paris) 170:856–858

Higuchi RG, Wilson AC (1984) Recovery of DNA from extinct species. Fed Proc 43:1557

Hunger R (1979) *The Magic of Amber*. Radnor, Penn.: Chilton Book Co.

Pääbo S (1990) Amplifying ancient DNA. In: Innis MA, Gelfand DH, Sninsky JJ, White TJ (eds) PCR Protocols: A Guide to Methods and Amplifications. San Diego: Academic Press, pp. 159–166

Poinar Jr GO (1982) Amber: true or false? Gems and Minerals 534:80–84

Poinar Jr GO (1992) *Life in Amber*, Palo Alto, Cal.: Stanford University Press

Poinar Jr GO, Hess R (1982) Ultrastructure of 40-million-year-old insect tissue. Science 215:1241–1242

Poinar Jr GO, Hess R (1985) Preservative qualities of recent and fossil resins: electron micrograph studies on tissue preserved in Baltic amber. J Baltic Studies 16:222–230

Poinar HN, Poinar Jr GO, Cano RJ (1993) Oldest DNA from plants. Nature 363:677.

Rice PC (1980) *Amber: The Golden Gem of the Ages*. New York: Van Nostrand Reinhold

Voordouw G, Voordouw JK, Karkhoff-Schweizer RR, Fedorak PM, Westlake DWS (1991) Reverse sample genome probing: a new technique for identification of bacteria in environmental samples by DNA hybridization, and its application to the identification of sulfate-reducing bacteria in oil field samples. Appl Environ Microbiol 57:3070–3078

Wet Samples
7 DNA Analysis of the Windover Population

WILLIAM W. HAUSWIRTH, CYNTHIA D. DICKEL, and DAVID A. LAWLOR

The discovery that a water-saturated environment can preserve DNA in tissue has provided a significant additional source of genetic material for the rapidly expanding field of ancient DNA (aDNA) analysis. In this chapter we describe some results of our genetic study of the Windover archaeological site. Windover is presently the most extensively characterized wet site with regard to ancient human DNA. Our intent is to describe the conditions of DNA preservation, discuss reasons for the surprisingly good state of that preservation, and outline the current status of genetic reconstruction of this 7,000–8,000-year-old population. Throughout, we have also attempted to highlight elements of our Windover study which may serve as useful predictive and methodological guides for retrieval and analysis of aDNA from other wet sites.

1. DNA Preservation at the Windover Site

The Windover site (8BR46) consists of a small (5,400 m^2) peat deposit in a low-lying swale on the western edge of the Florida Atlantic Coastal Ridge, roughly equidistant from the Indian River coastal lagoon system and the St. John's River in eastern central Florida. Information from preliminary analysis of flora and fauna indicates that the site was a wooded marsh from 8000 B.P. to 6900 B.P. and during this time was regularly used as a burial ground. A description and summary of the Windover site has appeared (Doran and Dickel 1988a, b). Most bodies found at the site had been placed in a flexed position and then buried lying on their sides in anaerobic, water-saturated peat at an approximate depth of 1 m. The chronometric placement of Windover skeletal material is based on a series of 14 radiocarbon dates. These dates were obtained directly from human bone, from the top and bottom of vertical burial stakes, and from peat above, within, and beneath human bone from multiple locations within the pond. The dates on human bone are consistent with all other dates and range from 6990 ± 70 to 8120 ± 70 years B.P., which would cluster human activities at Windover at approximately

7450 B.P. This places the utilization of Windover in a period that is usually considered Early Archaic in the southeastern United States (Milanich and Fairbanks 1980).

Over the period of excavation from 1984 to 1987, skeletal remains of 177 individuals of all ages were recovered. Associated with these remains were a large sample of prehistoric flexible fabrics and a more limited array of carved wooden and bone artifacts. Approximately half of the skeletal remains included intact crania, and within each of these, material visually and microscopically identifiable as preserved brain matter was found (Doran et al. 1986; Hauswirth et al. 1991). No other obviously preserved human tissue (except bone) was evident. An important parameter for the recovery of large amounts of soft tissue from wet sites therefore appears to be its confinement within a relatively impervious container, in this case the cranium, to prevent diffusion of the tissue into the surrounding matrix. Preliminary Southern blot analysis of small amounts of material associated with cranial fragments, recovered before intact crania were found, showed traces of human DNA (W.W. Hauswirth and P.J. Laipis, unpubl.) and confirms this conclusion. Additionally, although not specifically analyzed for DNA, the peat matrix immediately below many skeletal remains was discolored in the same way as brain matter within crania (D.N. Dickel, pers. comm.), suggesting that more sensitive techniques for DNA rescue (e.g., PCR) might have detected human DNA in this material as well. Thus, with molecular amplification techniques, intact tissue preservation at wet sites may not be necessary for genetic analysis and, given favorable chemical conditions (see below), many wet sites may contain useful levels of preserved tissue DNA in the matrix surrounding the skeletal remains.

The water chemistry of the Windover pond was clearly an important factor in DNA preservation; its general features may hold lessons for estimating the likelihood of comparable preservation at other sites. In contrast to a nearby pond, Windover water was highly mineralized (total dissolved solids 1,447 mg/l versus 170 mg/l for a neighboring pond; total hardness 716 mg/l versus 29 mg/l, with high amounts of calcium, magnesium, carbonates, sulfides, and sulfates). These conditions, coupled with an anaerobic environment present only a few centimeters below the surface, would have severely limited bacterial and fungal growth and subsequent tissue decomposition. Although anoxia itself does not totally inhibit bacterial decomposition of tissue (Allison 1988; Kidwellard and Baumiller 1990), this condition would have been important in limiting oxidative damage of DNA. Consistent with this conclusion, attempts to culture freshly disinterred samples of peat from within a cranium revealed only slow growth of microorganisms commonly found in soil and air.

Of particular relevance to preservation of DNA in wet tissue is the pH of the aqueous environment. The rate of acid-catalyzed depurination of DNA leading to strand breaks is very pH dependent (Lindahl and Nyberg 1972), and therefore DNA within tissue in typically acidic peat bogs and lakes (pH

< 4) should degrade rapidly. Only at pH values close to neutrality would analyzable amounts of DNA be expected to last several thousand years (Pääbo and Wilson 1991). Significantly, the Windover pond water is nearly neutral in pH (6.1 to 6.9), most likely due to the buffering capacity of carbonates derived from a snail shell layer found below the skeletal layer (Doran and Dickel 1988a) or from nearby limestone formations (Yamada et al. 1990). Thus a combination of several chemical features, including near-neutral pH, high mineralization, and low oxygen tension, appears to be critical in the wet site preservation of DNA. We suggest that where these conditions exist, the probability of rescuing informative amounts of DNA from preserved tissue is significantly enhanced.

2. Handling of Wet Site Tissue

Upon recovery of the first preserved brain mass from an intact cranium, the problems of handling wet site soft tissue became evident. Because there was interest in assessing the state of preservation of cellular structure, immediate freezing of the tissue for preservative storage was ruled out in our initial studies. Instead, the first five brain masses were removed from their crania, placed in sealable plastic bags, flushed with oxygen-free argon to inhibit oxidation, and stored sealed at 4°C. Once cellular analysis was complete (Doran et al. 1986), these and all other brain masses were stored sealed in argon at − 70° to − 80°C. After more than five years of such preservation for some samples, no reduction in either the total yield of human DNA or the quality of DNA as judged by its ability to serve as a PCR template has been observed.

The brain tissue, as recovered from each cranium, was structurally extremely fragile and crumbled easily if excessively handled. Therefore, to avoid harsh agitation of the tissue within a cranium and to minimize physical damage to the brain during removal from its cranium, X-ray images were routinely taken on the day of disinterment to locate the position of each brain mass. Although magnetic resonance imaging gave better resolution of brain versus peat with the cranium (Doran et al. 1986), it did not affect the decision of where to cut the cranium to most efficiently remove the brain mass, and was therefore not routinely performed. Throughout the handling of all brain material (even before the advent of PCR) surgical gloves were worn to minimize contamination from modern sources.

It is difficult to predict whether the lessons learned at Windover regarding handling of preserved soft matter will be generally applicable to other wet sites. However, the precautions employed at Windover and the tissue preservation technique described are simple, relatively inexpensive, and apparently effective. Their success at preserving aDNA relatively uncontaminated with modern material (see below) suggests that they should provide a starting point for similar efforts elsewhere. Specifically, we recommend: (1) handling all tissue, including bone if it is the intended source of aDNA, with surgical gloves and minimizing the time between disinterment and preservation; (2) removing oxygen from the environment of the sample as rapidly as possible by enclosure in heat-sealable plastic containers flushed with argon (if the samples cannot be frozen within a few days, repurging with argon will be necessary, since plastic bags are somewhat pervious to gases of small molecular diameter); (3) long-term preservation at $-70°$ to $-80°C$. Finally, the best-preserving condition for a wet site DNA sample is the original environment. Therefore, if possible, a portion of the ancient material should be left in situ, as was done at Windover for approximately 30% of the burial.

3. DNA Extraction and Pre-PCR Analysis

A detailed protocol for isolating DNA from preserved brain tissue has been published (Doran et al. 1986; Hauswirth et al. 1991); here we summarize the important features of that protocol. Nucleic acids were extracted and purified from 15 g of relatively peat-free brain cortex by homogenizing in 15 ml of 0.1 M EDTA, 0.15 M NaCl, 0.15 M sodium citrate at pH 7.6, 1% SDS with agitation for 16 hr. The solution was clarified by low-speed centrifugation, the supernatant chloroform/phenol-extracted and centrifuged in a CsCl–ethidium bromide density gradient. Material banding at a density of 1.55 g/cc was collected and identified as DNA by DNase sensitivity and RNase resistance. High molecular weight DNA of 8–20 kilobases (kb) was routinely found in all samples by ethidium bromide stained agarose gel analysis. Slight modifications of this technique are currently in use prior to PCR amplification (see Section 4).

To determine whether Windover DNA was of human origin, an agarose gel of this DNA was blotted and hybridized with a probe specific for human mitochondrial DNA (mtDNA) (Chang and Clayton 1982). The probe detected 16-kb open-circular and linear DNA in undigested brain DNA, demonstrating that human mtDNA was present (Doran et al. 1986). To confirm the presence of human nuclear DNA, a dot blot of ancient, modern, and peat DNA was hybridized with a cloned human *Alu* repeat sequence probe. Signal was obtained from the ancient and modern DNA, but not from the peat sample taken from the same stratigraphic level (Hauswirth et al. 1991). This experimental result was reproduced several times using DNA samples from different preserved brain tissues.

The total yield of DNA was about 1 µg/g tissue, or 1% of that normally isolated from fresh tissue. Curiously, the amount of mtDNA present in the ancient sample appeared to be low relative to the total amount of isolated DNA; a comparison of hybridization intensities between aDNA and modern human brain mtDNA when hybridized with the mtDNA probe suggested that only about 0.05% of the total aDNA was mtDNA, whereas DNA isolated from fresh brain tissue yielded 0.5–1% mtDNA. The nuclear DNA yield from preserved tissue was similarly reduced; quantitation of *Alu* sequences on dot blots estimated that human *Alu* sequences were present at only 1% of the level found in an equivalent amount of modern DNA (Hauswirth et al. 1991). Part of the reason for this became apparent when the surrounding peat was analyzed by gel electrophoresis and found to contain about the same amount of DNA on a per weight basis as the brain tissue. Importantly, however, peat DNA does not hybridize to either human mtDNA or human *Alu* DNA, and therefore its diffusion into brain tissue serves only to reduce the fraction of total DNA of human origin. Apparently conditions suitable to preserve human DNA will preserve plant DNA in the peat matrix as well. This lesson may well apply to other wet sites with similar conditions where plant material is present.

Further analysis revealed several other clues as to the chemical status of the aDNA. When the ancient mtDNA was digested with *Eco*RI, the 8-kb fragment which should appear after hybridization with the mtDNA probe was not present (Doran et al. 1986). However, partial conversion of open-circular to linear molecules did take place, as would be expected if only a fraction of the three *Eco*RI sequences in the mitochondrial genomes were being recognized by the restriction enzyme. Resistance to enzyme digestion was found to be an intrinsic property of the old DNA, because a bacterial plasmid DNA mixed in with aDNA did digest to completion under the same conditions. Thus Windover DNA contained nucleotide modifications leading to a loss of restriction enzyme sites. Based on the fraction of open-circular DNA converted to linear DNA by *Eco*RI we estimate that approximately one base in twenty was modified. The aDNA also lacked supercoiled, covalently closed mtDNA circles. The reason for their absence has not been investigated further, but it is known that many spontaneous processes can lead to single-strand nicks in DNA, converting covalently closed molecules to open-circular forms. Multiple single-strand nicks or damage resulting in a double-strand scission would lead to linear, full-length molecules; these were clearly evident.

High molecular weight DNA and a surprisingly high fraction of intact open-circular mtDNA was observed in all Windover DNA samples (Doran et al. 1986). Both of these observations may be due to DNA damage caused by depurination leading to interstrand crosslinking of DNA (Goffin et al. 1984). Such chemical crosslinking will raise the apparent double-strand molecular weight of the DNA, greatly increase the lifetime of circular DNA forms, and interfere with restriction digests. Crosslinking may also interfere with DNA hybridization experiments by preventing strand separation. This type of damage has been shown to be enhanced in aqueous solution (Goffin et al. 1984) and leads us to suggest that unlike some aDNA preserved in dry environments, wet site DNA may commonly exist at a surprisingly high molecular weight. However, molecular weight alone is not a valid indicator of DNA quality, as discussed above.

The presence and condition of nuclear DNA was also investigated. Restriction enzyme-digested DNA was Southern blotted and probed with radiolabeled probes for nuclear DNA and RNA sequences present in multiple copies in the human genome (a human *Alu* sequence and human large and small ribosomal RNAs) (Houck et al. 1979; Long and David 1980). None of these probes hybridized to restriction fragments of a defined size. If nuclear DNA was damaged in a similar manner as mtDNA and largely resistant to restriction digests, discrete restriction fragment bands of multicopy or single-copy genes would not occur in such hybridization experiments, consistent with our observations. In contrast, mtDNA occurs as a 16-kb circular molecule and is detectable as a discrete species independent of the recognition of specific undamaged sequences by a restriction endonuclease; therefore, its detection in hybridization experiments is expected. Alkaline cleavage of Windover DNA also revealed a significant amount of DNA damage (data not shown). We estimate that these alkali-sensitive sites, many of which are likely to represent apurinic deoxyribose phosphate sites, are present in about 1% of the nucleotides. To sum up, by a variety of analyses, 1–5% of the nucleotides in Windover DNA are damaged, which probably explains the difficulties in cloning or PCR-amplifying DNA fragments longer than a few hundred base pairs in length.

An attempt was made to directly clone Windover DNA from several samples of brain tissue. A library of DNA fragments was constructed using a partial *Alu*I digest and an M13 cloning vector (P.J. Laipis and W.W. Hauswirth, unpubl.). Approximately 1,000 clones containing small inserts (50–1,000 bp) were isolated, and 90 were screened for homology to human mtDNA, human *Alu* repeat sequences, or human ribosomal genes. The inserts from three clones exhibiting weakly positive hybridization signals were sequenced, but none exhibited similarity to any of the probes. Thus, although we were unable to confirm the presence of human DNA by direct cloning and sequencing, an occasional small fragment was cloneable, although at low efficiency (< 1% of the expected value). As an alternative approach to demonstrating the human origin of some cloned fragments of Windover DNA, we hybridized Southern blots of modern human brain DNA with probes made from the cloned ancient *Alu* DNA inserts described above. Twelve randomly selected probes were made and hybridized. In two instances discrete bands in modern human DNA were visualized, suggesting that at least a portion of the aDNA was of human nuclear origin. However, the precise identity of these cross-hybridizing sequences could not be determined by this technique because none of the probe sequences fortuitously matched a known human sequence. The development of PCR techniques preempted further attempts at direct cloning.

In summary, analysis of Windover DNA in the pre-PCR era, although unsuccessful at yielding informative DNA sequence information, provided the first definitive proof that ancient human DNA could be preserved in an aqueous environment and gave an estimate as to the frequency and kind of DNA damage occurring under such conditions. To our knowledge, Windover tissue has yielded the only successful Southern blot analysis for any species of DNA from an ancient source, with one possible exception (Rogan and

Salvo, 1990). This fact suggests (but does not prove) that wet sites may be particularly favorable environments for long-term DNA preservation.

4. PCR Protocols and Precautions

Polymerase chain reaction (PCR) has become the preferred molecular technique for characterization of DNA from ancient tissues. The sensitivity of the assay allows amplification of target sequences which may be at very low frequency in the DNA sample due to copy number or, in the case of the ancient samples, due to the presence of extensive chemical lesions in the substrate. Our current protocol for extraction of DNA from cerebral cortex tissue involves shaking 10 g tissue with 0.05 M EDTA, 0.3 M NaCl, 0.03 M sodium citrate, pH 7.4, and 1% SDS for 1 hr at 42°C. This is followed by two extractions with phenol/chloroform/isoamyl alcohol (25:24:1) and one extraction with chloroform. The aqueous phase is precipitated with 2.5 volumes of ethanol at $-20°C$ and the resultant precipitate redissolved in water. Cesium chloride (1 g/ml) and ethidium bromide (0.5 mg/ml) are added and the solution centrifuged to equilibrium in a TLA-100.3 rotor for 24 hr at 80,000 rpm in a Beckman TL-100 ultracentrifuge. Visible ethidium bromide-stained bands are collected, n-butanol extracted, and ethanol precipitated. The DNA is purified further by resuspension in 250 µl of H_2O followed by addition of 15 µl of 10 mg/ml ethidium bromide and 140 µl of 7.5 M ammonium acetate (Stemmer 1991). The solution is mixed well, then 420 µl of phenol/chloroform is added and centrifuged for 2 min. The upper, aqueous phase is removed and ethanol precipitated.

PCR is performed in 50-µl reaction volumes containing 1–10 ng DNA in 10 mM Tris-HCl, pH 8.3 (at 35°C); 50 mM KCl; 1.5 mM $MgCl_2$; 200 µM each of dATP, dCTP, dTTP, and dGTP; and 0.2 µM each of the 5' and 3' oligonucleotide primers. *Thermus aquaticus* (*Taq*) DNA polymerase (2 units) is added, and an initial denaturation of 94°C for 10 min is followed by 40 PCR cycles performed as follow: 94°C for 1 min, 2-min ramp time, 37°C for 1 min, 30-sec ramp time, 73°C for 3 min, and a 30-sec ramp time. Secondary amplification (booster PCR) is performed on a 1-µl aliquot of the completed reaction in a new PCR reaction containing fresh primers for 25 cycles during which the annealing temperature is 55°C.

The sensitivity of the PCR assay also provides its greatest potential pitfall: the detection and amplification of contaminating DNA. The most likely contaminant is modern human DNA, although cross-contamination with other aDNA samples is theoretically possible. We only mention precautions specific for work done in our laboratories. For the analysis of MHC class I genes (see below; Lawlor et al. 1991), the work is divided between laboratories in order to minimize amplification of contaminant molecules. The amplification of aDNA is performed at the University of Florida laboratory because MHC class I clones have not been handled at this facility, while the cloning and sequencing of the amplified product is conducted at Houston (previously at Stanford). Also, the amplification primers utilized for MHC class I genes are synthesized and reserved for use at the Florida laboratory; no PCR experiments with these specific primers are performed at the other laboratories. The personnel performing the experiments have been tissue typed to determine their HLA-A, -B, and -C specificities; their types are dissimilar to all those currently identified in the Windover individuals. PCR reactions of aDNA are assembled in a laminar flow hood equipped with a UV light source. No modern DNA is introduced into this area at any time. Dedicated pipets are used, and gloves are worn at all times. Pipet tips with filter barriers are used for all solutions. External contamination is essentially eliminated by these precautions. Any cross-contamination still present is reduced further by minimal handling of PCR reaction tubes and use of tube openers instead of gloved hands to open tubes. PCR reaction controls containing all reagents

TABLE 1. Windover class I MHC and β_2-microglobulin clones

Individual and clone[1]	PCR experiment	# Positive/ total[2]/	Gene designation
SS325-5	1	19/23	DAN4
SS325-1	1	1/23	HLA-G
SS325-2	1	1/23	HLA-G-like
SS325-3	1	1/23	HLA-G + PatrB2
SS325-4	1	1/23	HLA-H + B*1501
SS325-10	2	12/19	B*3701
SS325-8	2	3/19	DAN4
SS325-6	2	1/19	PatrB2-like
SS325-7	2	1/19	PatrB2-like
SS325-11	2	1/19	B*3701 + A*0201
SS325-12	2	1/19	B*3701 + A*0201 + B*3701
SS325-14	4	43/45	A19 (A*2901, A*3101 or A*3201)
SS325-β_2m	5	11/11	β_2-microglobulin
SS58-1	6	7/7	DAN2

[1] DNA from two individuals, SS325 and SS58, was amplified with MHC class I primers (5'-ATC GAA GCT TAA GGA TTA CAT CGC CCT GAA CGA GGA-3' and 5'-AGT CGT CGA CCT CCA GGT ATC TGC GGA GCC ACT CCA-3') or β_{-2}microglobulin primers (5'-CCG CAA GCT TTC TCC ATT CTT CAG TAA GTC AAC TTC-3' and 5'-GGG CGT CGA CAT TCA GGT TTA CTC ACG TCA TCC AGC-3') and the product was cloned into M13 for DNA sequence analysis.

[2] Number of independent clones with the respective sequence and total number of MHC-containing clones obtained from the PCR experiment.

[3] Many of the PCR clones contained sequences identical to modern-day MHC class I genes or β_2-microglobulin. Several of the clones, such as SS325-2, differed from a corresponding gene by one or two substitutions, possibly a result of polymerase error. Others, such as SS325-12, are chimeric products containing segments from two or more class I genes.

except DNA are used to detect any external cross-contamination. In our analysis of MHC class I genes from SS325 DNA, multiple experiments were performed in order to demonstrate the retrieval of the same sequence from independent extractions. SS325-5 and SS325-8 have identical sequences (Table 1 and Section 5) but were obtained from separate extractions and PCR of independent ancient tissue. This precaution provides insurance of authenticity in cases where a possible contaminant is introduced during the DNA extraction procedure.

5. Analysis of Single-Copy Nuclear Genes: The MHC Class I and β_2-Microglobulin Genes

The major histocompatibility complex (MHC) in humans contains 17 to 20 class I genes as assessed by Southern blot analysis. The genes for which DNA sequence information exists are greater than 80% similar. The family contains genes (HLA-A, -B, -C) that are transcribed in nearly all cell types and encode molecules which bind and transport internally processed peptide fragments to the cell surface. These genes exhibit an extraordinary degree of diversification, with multiple alleles existing in low but appreciable frequencies within the human population. Presently more than 100 alleles have been characterized; some estimates would place the total number at double that figure. The explosive increase in allelic characterizations is due to the employment of molecular techniques with greater sensitivity for detecting serologically silent subtypes and variants. Although the majority of HLA-A, -B, and -C alleles occur at low frequency, there are exceptions such as HLA-A2 which is present in 45% of the Caucasian population. Also, some alleles are rare in certain racial groups and present in much higher frequency in others. Examples include HLA-A*3401 and HLA-B*4201, which occur in fewer than 1% of Caucasians but in more than 15% of Africans. Other members of the MHC class I gene family (HLA-E, -F, -G) are differentially transcribed in various tissues; their immune function remains unclear. Although there is only partial characterization of these genes, it is clear that locus diversification is far less than for A, B, and C. There are five HLA-E alleles and fewer yet for G and F. Finally, the MHC-I family also contains pseudogenes, such as HLA-H and DAN4, with mutations that preclude productive transcription.

Diversification of the A, B, and C loci results primarily from positive selection for replacement substitutions at codons comprising the antigen binding site of these molecules. The codons are located in exons encoding the membrane-distal domains of this cell surface molecule. Much of the variability which distinguishes loci as well as alleles of a locus is localized in short stretches within these domains of the gene. The variable regions are flanked by relatively invariant segments which are conserved in all class I genes characterized to date. This allowed us to design amplification primers which would bind to invariant regions yet amplify highly diagnostic target sequences from the majority, if not all, class I genes (Lawlor et al. 1991). Three pairs of heavy-chain primers were designed after assessment of a large data base of MHC class I sequences. A fourth pair of primers was designed to amplify a 62-bp stretch of β_2 microglobulin, a single-copy gene with no allelic variation. Sizes of the target sequences were small, the largest being 125 bp. All four pairs efficiently amplified modern DNA as assessed by agarose gel electrophoresis, but only pair #3 and the β_2 microglobulin primers successfully produced product of the desired size when Windover DNA served as the PCR substrate. The reason(s) for the failure of pairs #1 and 2 to amplify the aDNA are unclear. Other approaches are now being employed to derive sequence information from those regions. However, this result supports the connection that the aDNA was highly damaged and devoid of modern DNA contaminants.

For sequence analysis, PCR products with pair #3 and the β_2-microglobulin primers were digested with *Sal*I and *Hind*III, as these sites had been added to the primers, followed by cloning into an M13 vector. Direct sequencing of amplified product may be preferable when the target DNA segment derives from a single gene; however, our class I heavy-chain primers were designed to amplify all members of a large family of related genes, and direct sequencing was not a viable approach. To eliminate errors due to *Taq* polymerase misincorporations or postmortem modifications of the DNA, we routinely sequenced multiple clones to ensure accurate determination. Since the β_2 microglobulin gene is relatively invariant, the faithful retrieval of this sequence from Windoever DNA documented the usefulness of PCR for analysis. For primer pair #3, the background of irrelevant sequences was reduced substantially by secondary amplification with a slightly indented pair of primers. This technique increased the number of class I-containing clones to above 90%. Although there was a slight decrease in target size (107 bp instead of 125) the dramatic enrichment for the desired target sequence justified the incorporation of internal primers for secondary amplifications.

Two prominent features of aDNA PCR both related to the distribution of cloned PCR products. In successful amplifications, a single sequence usually predominated and constituted 40–100% of the positive clones. The frequency and the nature of the sequence varied between amplifications (Table 1). The most extreme case occurred when 43 of the 45 class-I-containing clones contained a sequence identical to HLA-A19. The remaining two clones had single base pair substitutions from the HLA-A19 sequence, possibly the result of polymerase errors. In all cases the predominant sequence was invariably a faithful copy of a segment derived from a modern HLA allele. In our first amplification, 19 of 23 positive clones contained a sequence which did not correspond to any class I gene in our data base at that time. The sequence was only 80% similar to any class I gene and contained 124 bp instead of the expected 125. These data suggested that the sequence was derived from an uncharacterized pseudogene in the class I family. Validation of the sequence was provided by independent subsequent characterization of the class I gene, DAN4, from a modern cell line (Hansen et al. 1991). Our sequence and DAN4 were identical (over the length of comparison) and the modern sequence had the same single base pair deletion that characterized the aDNA.

In contrast to the sequence fidelity of the predominant clones, minority products were often chimeras, consisting of segments derived from more than one class I gene. The formation of a chimeric product is thought to result from incomplete chain extension during PCR (termed "template jumping" by Pääbo et al. 1990). The incomplete strands can reanneal in later cycles to other class I templates, with subsequent formation of chimeric sequences. The presence of chimeric sequences suggests that the DNA substrate is damaged, as expected for aDNA. Chimeras are always a minority species since several cycles of amplification are required to span the two priming sites. Thus chimeric products accumulate in an additive fashion until priming sites are bridged, at which time they start to accumulate exponentially. In contrast, exponential amplification of intact segments starts with the first cycle. Although chimeras are clearly in vitro artifacts, they are not uninformative as the nature of donor genes often can be inferred from the hybrid sequences.

With the analysis of four amplification of SS325 DNA, it was possible to define the HLA-A and -B alleles possessed by this adult male. At the A locus, this person expressed an allele of the A19 family, either A*2901, A*3101, or A*3201. Most likely it was HLA-A*3101, an allele which is found frequently in modern American Indians. Final resolution will require retrieval of sequence information from one of the other two target segments. Examination of the chimeric sequences provides suggestive evidence that HLA-A*0201 was the other A locus allele. SS325 was also heterozygous at the B locus, possessing B*3701 and an allele related to but not identical with B*1801 or B*1501. HLA-B*3701 is an allele that occurs in low frequency in most racial groups (1.5% in Caucasians), but there is little information concerning its frequency in American Indian populations. The second B locus allele has not yet been described in modern populations. Although a complete canvassing of B locus alleles may uncover this sequence, the possibility exists that the allele no longer exists in the modern human gene pool. Other class I genes which have been identified in SS325, or other Windover individuals, include HLA-G, HLA-H, DAN4, and DAN2. In sum, this analysis demonstrated for the first time the possibility of nuclear DNA analysis from aDNA.

6. Mitochondrial DNA Sequence Analysis

As described in Section 3, Southern blot analysis of Windover DNA demonstrated the presence of human mtDNA at low but analyzable levels. Direct cloning and sequence analysis failed to confirm this result, probably because the integrity of the DNA was poor, with at least 1–5% of the nucleotides being chemically altered. Thus, few if any mtDNA sequences greater than

200–300 bp would be suitable templates for PCR analysis. With this limitation in mind, we targeted PCR sequence analysis to the 3′ end of the human mitochondrial d-loop region because of its well-documented hypervariability (Aquadro and Greenberg 1983) and its utility in estimating the history of maternal lineages in a population (Ward et al. 1991). Our initial aim was to establish the level of genetic diversity in the Windover population. PCR primer pairs were targeted to amplify the region between 16,160 and 16,320 in the mitochondrial genome (Table 2). Primary PCR products were resolved in low melting temperature agarose gels and the appropriate band excised. Two to three μl of the melted 50-μl gel fragments was then used in a second PCR reaction to which one primer was added at $\frac{1}{50}$ of the normal concentration in order to make predominantly single-stranded PCR products for standard dideoxy-chain termination sequencing. The results for seven Windover samples are shown in Table 2.

Several features of the d-loop sequence comparison are worth noting. Five of the seven individuals tested had the identical sequence. The implications of this are discussed below. The other two individuals, SS520 and SS108, each had three single-base differences compared to the predominant genotype. None of the six sites distinguishing these two individuals are coincident, leading to the tentative conclusion that the three mitochondrial genotypes found thus far are unrelated, at least in the sense that they do not form an obvious set of sequences evolving stepwise from one genotype to another. Several characteristics of the sequences argue against an illegitimate origin due to PCR artifact. Five of the six differences are transition substitutions, as expected among relatively closely related mitochondrial sequences (Miyata et al. 1982) and unlike the more equal rate of transition and transversion substitutions expected for PCR-generated sequence variation (S. Pääbo, pers. comm.). Additionally, because sequences were determined directly from total PCR product DNA and not from cloned molecules as was necessary for the MHC class I gene analysis, there was no reason that a single aberrant sequence would predominate, unless it was generated in the first few PCR cycles. If so, multiple independent PCR sequence determinations would yield different results, and this was not the case; each of the two variant sequences was redetermined from another primary PCR step and found to be accurate.

Although it is probably too early in this analysis to draw any firm conclusions, because of the limited number of DNA samples and the limited d-loop domain analyzed, the trends thus far are intriguing. That five of the seven DNAs are identical suggests homogeneity in the maternal lineages of Windover. This is very different from the approximately twofold higher level of intrapopulation diversity noted in a modern African pigmy tribe (Vigilant et al. 1991) and in a modern northwestern American Indian group (Ward et al. 1991). Possible reasons might relate to a more limited number of founding maternal lineages for Windover, a genetic bottleneck due to severe population reduction occurring in the Windover population just before or during the burial period (7000–8000 years B.P.), less mobility of this small, interrelated

TABLE 2. Mitochondrial d-loop sequence analysis of Windover individuals

	16220		16240			16260
	CAAGCAAGTA	CAGCAATCAA	CCCTCAACTA	TCACACATCA	ACTGCAACTC	CAAAGCCACC
325	----------	----------	----------	----------	----------	----------
520	---T------	----------	--C-------	----------	-T--------	--T-------
108	----------	----------	----------	----------	----------	----------
61	----------	----------	----------	----------	----------	----------
217	----------	----------	----------	----------	----------	----------
199	----------	----------	----------	----------	----------	----------
58	----------	----------	----------	----------	----------	----------

	16280		16300			16320
	CCTCACCCAC	TAGGATACCA	ACAAACCTAC	CCACCCTTAA	CAGTACATAG	TACATAAAGG
325	----------	----------	----------	----------	----------	----------
520	---T------	----------	----------	----------	----------	----------
108	--------T-	----------	----------	----------	----------	----------
61	----------	----------	----------	----------	----------	----------
217	----------	----------	----------	----------	----------	----------
199	----------	----------	----------	----------	----------	----------
58	----------	----------	----------	----------	----------	----------

A portion of the hypervariable 3′ d-loop is shown for seven different Windover brain DNAs (left). Only differences from the reference sequence (sample 325) are indicated. The PCR primers were 5′-TAC TTG ACC ACC TGT AGT AC-3′ and 5′-CCG CGA ATT CTG TGC TAT GTA CGG TAA ATG-3′.

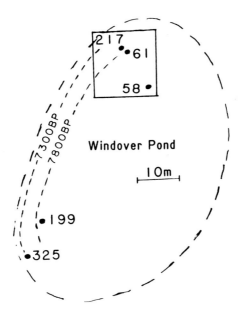

FIGURE 1. Location of mitochondrial d-loop haplotypes at the Windover archaeo-logical site. The five Windover brain DNA samples exhibiting the identical sequence in the targeted hypervariable d-loop region (Table 2) are shown relative to the Windover pond margin (– –). Additionally, based on current radiocarbon data, approximate age contours are shown (— —).

Windover population and less genetic contact with outside maternal lineages, or a combination of these factors.

Because burials at Windover are likely to have occurred over an extended period of time (about 1000 years, based on current radiocarbon data) it is possible to estimate the continuity of maternal lineages. This is a unique genetic parameter not easily obtainable by studying living populations. One way to approach this question at Windover is to ask whether individuals with the same mitochondrial genotype were interred at divergent locales within the Windover pond burial. This is particularly relevant since radio-carbon dates suggest a possible temporal pattern to present burial locations of samples. Figure 1 depicts the Windover pond and the locales of the burials (G.H. Doran, pers. comm.) containing the predominant mitochondrial sequence (haplotype A). Two features are notable. First, the locations of mitochondrial haplotype A individual span the extremes of burial locations (325 and 199 relative to 217, 61, and 58). Second, these burial sites also span the extremes of distance from the current pond margin (325 relative to 58). It is this latter parameter which appears to correlate with radiocarbon age; burials near the pond edge are about 7,000 years old, those farthest from the edge are more than 8,000 years old. More samples are currently being radiocarbon dated to substantiate this feature of the site, but it appears that at least one mitochondrial haplotype spans the entire burial period of Windover, approximately 50 human generations. In summary, mtDNA sequence analysis demonstrate the viability of ancient wet site DNA for genetic analysis and portends widespread utility in ancient population studies.

7. Conclusions and Prospects

The Windover site has yielded preserved brain tissue of human origin dating to the Early Archaic Period in North America. It is the oldest human soft tissue yet analyzed at the molecular level. PCR amplification, DNA cloning, and sequence analysis have demonstrated the presence of single-copy human nuclear genes representing alleles of the MHC class I gene complex on human chromosome 6 and human mtDNA. Although only a limited number of samples have been analyzed so far, it appears that the Windover population may exhibit a surprising degree of genetic homogeneity. These observations raise questions in two broad categories: first, what factors resulted in the preservation at Windover and second, what type of information does the presence of ancient nuclear and mtDNA make available to molecular, evolutionary, and population biology and other disciplines.

Upon comparing descriptions of preserved tissue found at other archaeological or forensic sites with the material found at Windover, we note that specimens most similar to Windover came from sites where adipocere formation occurred. Adipocere, also known as "grave wax," is a mixture of free fatty acids, primarily palmitic acid, and soaps resulting from postmortem hydrolysis and hydrogenation of naturally occurring fat in body tissue (Mant 1957; Zivanovic 1982). It requires the presence of aqueous conditions, fatty tissue, electrolytes (which may come from body fluids), and some putrification (to initiate the process). Apparently at Windover, burial practices as well as physical and chemical conditions allowed putrification to begin, but the process was halted before complete decomposition of brain tissue occurred.

Preserved human tissue of this antiquity from non-acidic, water-saturated environments like Windover is unusual but not unique. Human skulls containing the apparent remnants of brain tissue have previously been found at several Florida sites (Royal and Clark 1960; Dailey et al. 1972; Clausen et al. 1979; Beriault et al. 1981). A Danish medieval cemetery yielded 56 skulls with brain material (Tkocz et al. 1979); as at the Windover site, no other soft tissue was preserved. However, the great majority of bog bodies of northern Europe are from acidic environments (pH < 4), although some have been found under less acidic conditions (pH 5–7.5) with adipocere present (Fischer 1980). Clearly, in the appropriate natural environment, soft tissue and presumably genetic material can be preserved for a long period of time. The finds at Windover should encourage re-examination of all these sites.

The conditions at Windover which most likely inhibit bacterial growth and therefore contribute to enhanced tissue preservation are the high amounts of minerals present in the water and the anoxic conditions which begin just below the peat surface. DNA within the tissue was preserved due to at least two factors: first, water at Windover is nearly neutral (pH 6.1–6.9) and second, the anaerobic property of the water has limited oxidative DNA damage. Thus, DNA alteration due to acid depurination, deamination, and oxidation was minimized. Interestingly, low temperature may not have been a factor in DNA preservation since present-day subsurface groundwater temperatures are around 23°C and would not have been substantially different in 8000 B.P. Adipocere formation is indeed possible under warm water conditions, as was previously noted when two human bodies immersed in 21°C water for 5 years were recovered with extensive adipocere present in the bodies (Cotton et al. 1987). Although DNA has been isolated from tissues preserved through rapid drying (see section on Dried Samples), our results demonstrate that tissues recovered from water-saturated environments under conditions of limited oxygen, neutral pH, and high ion levels also yield preserved DNA of acceptable quality.

In an archaeological setting, DNA survival depends not only on chemical factors but also on ethnological practices influencing burial conditions and, thus, rate of tissue decomposition and DNA survival. Individual differences in tissue integrity between burial sites may reflect variations in either burial environment or the interval and conditions which prevailed between death and interment, or both. Preservation may also reflect sex- and age-specific burial patterns. However, at Windover, brain material has been recovered from infants, adolescents, and young and old adults representing both sexes, indicating little status distinction in burial patterns (Doran and Dickel 1988a). Agreement in age between the peat surrounding the skeletal material and the bone itself suggests primary burials in shallow graves. There are some indications that the bodies may have been deposited in water deep enough to require pointed "hold down" stakes. We suggest, therefore, that in the temperature latitudes of the New World practices involving rapid, simple burials in an anaerobic, water-saturated matrix may be an important factor in soft tissue DNA preservation.

At the genetic level, the prospect of HLA typing the Windover population and determining maternal lineages through mtDNA d-loop sequence analysis portends rapid advances in our understanding to the origins and migrations of ancient American populations, as well as contributing to a more basic understanding of MHC class I function and variation and the mechanisms of mtDNA change.

Acknowledgments. The authors thank Drs. Glen Doran and David Dickel for their generosity in sharing data, their professional oversight of the Windover archaeological site, and their continuing collaboration on all anthropological aspects of Windover; Dr. Philip Laipis for his collaboration on the initial analysis of Windover DNA; Dr. Peter Parham for his advice and support of one of us (D.A.L.); Sandra Doyle for her timely and competent word processing. We acknowledge partial support from NIH GM34825 and the Society to Prevent Blindness, Inc.

References

Allison PA (1988) The role of anoxia in the decay and mineralization of proteinaceous macro-fossils. Paleobiology 14:139–154

Aquadro CF, Greenberg BD (1983) Human mitochondrial DNA variation and evolution: analysis of nucleotide sequences from seven individuals. Genetics 103:287–312

Beriault J, Carr R, Stipp J, Johnson R and Meeder J (1981) The archaeological salvage of the Bay West Site, Collier County, Florida. Florida Anthrop 34:39–58

Chang DD, Clayton DA (1985) Priming of human mitochondrial DNA replication occurs at the light strand promoter. Proc Natl Acad Sci USA 82:351–355

Clausen CJ, Cohen AD, Emiliani C, Holman JA, Stipp JJ (1979) Little Salt Spring, Florida: a unique underwater site. Science 203:609–614

Cotton GE, Auferheide AC, Goldschmidt VG (1987) Preservation of human tissue immersed for five years in fresh water of known temperature. J Forensic Sci 32:1125–1130

Dailey RC, Morrell LR, Cockrell WA (1972) The St. Marks Cemetery (8WA108). Bureau of Historic Sites and Properties Special Reports 2:1–24

Doran GH, Dickel DN (1988a) Multidisciplinary investigations at the Windover Site In: Purdy B (ed) *Wet Site Archeology.* Caldwell, N.J.: Telford Press, pp. 263–289

Doran GH, Dickel DN (1988b) Radiometric chronology of the Archaic Windover archeological site. Florida Anthrop 41:365–380

Doran G, Dickel D, Agee FO, Ballinger WA, Laipis PJ, Hauswirth WW (1986) 8000 year old human brain tissue: anatomical, cellular and molecular analysis. Nature 323:803–806

Fischer C (1980) Bog bodies of Denmark. In: Cockburn AC, Cockburn E (eds) *Mummies, Disease and Ancient Cultures.* Cambridge, U.K.: Cambridge University Press, pp. 177–193

Goffin C, Bricteux-Gregoire S, Verly WG (1984) Some properties of the interstrand crosslinks in depurinated DNA. Biochem Biophys Acta 783:1–5

Hansen T, Markussen G, Paulsen G, Thorsby E (1991) Partial DNA sequences derived from two previously unknown HLA class I genes. Tissue Antigens 37:16–20

Hauswirth WW, Dickel CD, Doran GH, Laipis PJ, Dickel DN (1991) 8,000 year old human brain tissue from the Windover site: anatomical, cellular and molecular analysis. In: Ortner DJ, Aufderheide AC (eds) *Human Paleopathology: Current Synthesis and Future Options* Washington, D.C.: Smithsonian Press, pp. 60–72

Houck CM, Rinehart FP, Schmid CW (1979) A ubiquitous family of repeated DNA sequences in the human genome. J Mol Biol 132:289–306

Kidwell SM, Baumiller T (1990) Experimental disintegration of regular echinoids: roles of temperature, oxygen, and decay thresholds. Paleobiology 16:247–271

Lawlor DA, Dickel CD, Hauswirth WW, Parham P (1991) Ancient HLA from 7500 year old archaeological remains. Nature 349:785–788

Lindahl T, Nyberg B (1972) Rate of depurination of native deoxyribonucleic acid. Biochemistry 11:3610–3618

Long EO, Dawid IB (1980) Repeated genes in eucaryotes. Annu Rev Biochem 49:727–764

Mant AK (1957) Adipocere: a review. J Forensic Med 4:18–35

Milanich J, Fairbanks C (1980) *Florida Archeology*. New York: Academic Press

Miyata T, Hayashida H, Kikuno R, Hasegawa M, Kobayashi M, Koike K (1982) Molecular clock of silent substitution: at least a six-fold preponderance of silent changes in mitochondrial genes over those in nuclear genes. J Mol Evol 19:28–35

Pääbo S, Wilson AC (1991) Molecular evolution: Miocene DNA sequences—a dream come true? Curr Biol 1:45–46

Pääbo S, Irwin DM, Wilson AC (1990) Enzymatic amplification from modified DNA templates. J Biol Chem 265:4718–4721

Rogan PK, Salvo JJ (1990) Molecular genetics of pre-Columbian South American mummies. Yearbk Phys Anthropol 33:195–206

Royal W, Clark E (1960) Natural preservation of human brain, Warm Mineral Springs, Florida. Antiquity 26:285–287

Stemmer WPC (1991) A 20-minute ethidium bromide/high-salt extraction protocol for plasmid DNA. Biotechnology 10:726

Tkocz I, Bytzer P, Bierring F (1979) Preserved brains in medieval skulls. Am J Phys Anthrop 51:197–202

Vigilant L, Stoneking M, Harpending H, Hawkes K, Wilson AC (1991) African populations and the evolution of human mitochondrial DNA. Science 253:1503–1507

Ward RH, Frazier BS, Dew K, Pääbo S (1991) Extensive mitochondrial diversity within a single American Indian tribe. Proc Natl Acad Sci USA 88:8720–8724

Yamada TK, Kudou T, Takahashi-Iwanaga H (1990) Some 320-year-old soft tissue preserved by the presence of mercury. J Archaelog Sci 17:383–392

Zivanovic S (1982) *Ancient Diseases: The Elements of Paleopathology*. London: Methuen

ago. The body was extraordinarily well preserved and the find has triggered the organization of an international multidisciplinary research program including DNA studies (cf. Coghlan 1992).

In conclusion, these finds are of very different kinds, and their exploration may challenge our current knowledge about aDNA in many ways. Our starting point for the studies to be described below has been problems related to the history of Greenland. Greenland is a well-defined geographical and cultural entity, but many questions relating to its colonization and decolonization by the Eskimos and the Norse are unsolved (see Section 4 below). Considering the lack of written records, new approaches such as DNA studies are highly pertinent. So far, we have focused on the analysis of one very important find, the Qilakitsoq mummies.

2. The Qilakitsoq Mummies

In 1972, two burials were found in a rock cleft near the abandoned settlement Qilakitsoq close to Uummannaq in western Greenland (70 40′ N, 52 0′ W). The burials were excavated in 1978 and contained five and three mummified bodies, respectively. The find was radiocarbon dated to A.D. 1475 ± 50 years. The bodies were remarkably well preserved as a result of favorable local conditions. The burials were located on a slope facing north and protected from direct rain and snow by an overhanging rock. Mummification took place by desiccation at low temperatures which stopped normal postmortem decomposition. The burials were well aerated and the temperature inside them probably never exceeded 5°C. The Qilakitsoq find has been subject to a multidisciplinary study involving researchers in natural science, medicine, and cultural history. The provisional results of this study were compiled by Hart Hansen and Gulløv (1989). At a later stage, an interest in studying the DNA of these mummies emerged. The four best-preserved mummies were conserved in toto by 2–2.5 megarad of gamma radiation and are therefore probably useless for DNA studies. However, samples were available from one of the poorly preserved mummies labeled I/5 (Fig. 1). The soft tissues of this body had disappeared or were found as uncharacteristic brittle, and the body was partly skeletonized. White mold was encountered on all surfaces. I/5 was a woman 40–50 years old at the time of death. She was fully dressed in skin garments and wrapped in a caribou skin. She was found lying in the bottom of the vertical grave on top of plant material and separated from the above-lying body of I/4 by several layers of skins. Her position at the bottom of the grave probably was a contributing factor to the relatively poor preservation of the mummy. No special precautions were taken during excavation and subsequent handling to avoid contamination of the mummy or to facilitate the recovery of DNA. The mummy was stored at room temperature for more than 10 years before DNA studies were initiated. The exact history of the samples used in the present study is unknown.

FIGURE 1. Mummy labeled 1/5 from Qilakitsoq, Greenland. According to anthropological evidence this mummy was a 40–50-year-old woman. The find was radiocarbon dated to 1475 ± 50 years. (Photograph: Lennart Larsen, Danish National Museum.)

3. Characterization of DNA from Skin and Bone Samples

In a previous publication we reported on the isolation of human DNA from skin and bone samples from the Qilakitsoq mummy I/5 (Thuesen and Engberg 1990). This work is summarized here, and the result is compared to that of a new study involving three different DNA extraction protocols applied to a bone sample from the same mummy. In addition, we report on the further characterization of the DNA described in our original paper. The polymerase chain reaction (PCR) method was used to determine the sex of the mummy and to generate material from mitochondrial DNA (mtDNA) for sequence analysis in order to study its ethnic origin.

3.1 DNA Extraction and Purification

The skin sample consisted of epidermis, dermis, and subcutis. The skin was dry and shrunken, dark brown, and had a woodlike texture. About 5 g of this material was finely cut with a pair of scissors in preparation for DNA extraction. The bone sample was from sternum and clavicula. The bones were cleaned by rinsing with water and ground up with a pair of cutting pliers. Bone and skin samples were mixed with 25 ml of 60°C extraction buffer (10 mM Tris-HCl, 0.5 M EDTA, pH 9.5, and 1% SDS) after which 10 mg of proteinase K (Boehringer Mannheim) and 0.5 mg of *E. coli* tRNA (Sigma) was added. After incubation for 30 min at 60°C, the mixture was left overnight at 37°C with gentle shaking. The skin extract became intensely dark-colored during this incubation. It was cleared from debris by centrifugation at $10,000 \times g$ for 10 min. The pelleted material was extracted a second time with 20 ml of extraction buffer, centrifuged, and pooled with the first extract. Skin and bone extracts were then extracted five times with phenol/chloroform (1:1) and the nucleic acids collected by ethanol precipitation. The material was further purified by gel filtration through a Sephadex column (G-50). At this stage the skin extract had lost most of the dark-colored material. The eluate from the gel filtration column constituted the material used for the DNA analyses described in the following sections.

In a more recent experiment we set out to compare three different DNA extraction protocols all applied to a bone sample from the Qilakitsoq mummy I/5. For this experiment we chose a piece of femur. This bone was not as brittle as those used in the study described above, and a different approach had to be taken in order to prepare the sample for DNA extraction. The bone was split longitudinally and sawed into slices of about 1 g each. The slices were frozen in liquid nitrogen and ground up in a freeze mill. The resulting powder was subjected to extraction according to one of the following protocols.

The first extraction protocol was similar to our original protocol as described above up to the point of phenol extraction, except that only 10 ml of extraction buffer was used and the pH of the buffer was lowered from 9.5 to 8.0. The second protocol was taken from Pääbo (1990). The bone powder was mixed with 10 ml of extraction buffer (10 mM Tris-HCl, pH 8.0, 2 mM EDTA, 10 mg/ml DTT, 0.5 mg/ml proteinase K, 0.1% SDS) and incubated overnight at 37°C. The third extraction protocol was based on the following line of reasoning. Living, adult bone consists of about 70% inorganic mineral distributed throughout an organic matrix (Sillen 1989). The bulk of the inorganic phase is composed of crystalline and amorphous calcium phosphate. The crystalline calcium phosphate is largely hydroxyapatite ($Ca_5(PO_4)_3OH$) in the form of relatively small crystals. Although the composition of the calcium phosphates in bone is subject to changes during decomposition, it is likely that the hydroxyapatite content is high in mummified bones because they have been protected from exposure to acidic solutions. Hydroxyapatite is used in molecular biology as an affinity medium for nucleic acids. The nucleic acids bind to hydroxyapatite by virtue of interactions between the phosphate groups of the polynucleotide backbone and calcium residues in the resin. Bound nucleic acids can be eluted in phosphate buffers at elevated temperature. Double-stranded DNA binds more tightly to hydroxyapatite than does single-stranded DNA, and it requires higher phosphate concentrations for elution (for a general description of the use of hydroxyapatite, see Sambrook et al. 1989). We speculated that some of the DNA in bones will become bound to the hydroxyapatite crystals during decomposition. Moreover, it is possible that this bound DNA is protected from chemical and microbial degradation. In order to test this idea, we mixed the bone powder with 10 ml of 0.5 M sodium phosphate (pH 6.8) and incubated the sample overnight a 60°C prior to the phenol extraction step. This entire procedure is referred to as the third extraction protocol.

As a control for the three extraction protocols, we performed parallel "mock extractions" of the extraction buffer. All samples were centrifuged and the supernatants extracted twice with one volume of buffered phenol/chloroform (9:1). Aliquots of 2 ml of the resulting aqueous phases were then concentrated and purified to remove small molecular weight compounds by ultrafiltration using Centricon 30 microconcentrators (Amicon Inc.). Two consecutive cycles of dilution with distilled water and subsequent concentration were used and the material was finally recovered in a volume of 100 μl.

A gel electrophoretic analysis of the extracts is depicted in Figure 2. A uniform smear of DNA representing fragments ranging in size from less than 100 bp to more than 10 kbp is seen in lanes 1–4, whereas lanes 5–7 containing "mock extracts" are blank. This indicates that the three extraction protocols resulted in DNA of comparable size distributions. The DNA yield was highest when the protocol modified from Thuesen and Engberg (1990) was followed. The extraction protocol adopted from Pääbo (1990) gave approximately half this amount of DNA, whereas our new protocol produced much less. It is possible that the success of the Thuesen and Engberg (1990) protocol is due to decalcification of the bone powder by the high EDTA concentration. The

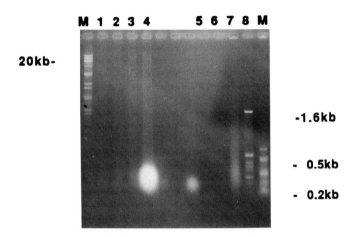

FIGURE 2. Gel electrophoretic analysis of bone extracts. 10/100 µl of each extract was mixed with gel loading buffer and run on a 1.4% agarose gel in TAE-buffer, along with DNA molecular weight markers. Following electrophoresis, the gel was stained with ethidium bromide and photographed in a UV transilluminator (for details of general methods, see Sambrook et al. 1989). Lanes 1, 2: DNA eluted and extracted as if bound to hydroxyapatite (see text for details). The phenol extraction step included the bone powder (lane 2) or was of the aqueous phase only (lane 1). Lane 3: extraction according to Pääbo (1990). Lane 4: extraction modified from Thuesen and Engberg (1990). Lanes 5–7: mock extracts corresponding to lanes 2–4. Lane 8: extract from Thuesen and Engberg (1990). M: molecular weight markers. A few selected fragment sizes are indicated.

DNA in lane 8 (Fig. 2) is taken from our original study. The mean size of this DNA is less than 500 bp, and very little DNA exceeds a fragment size of more than a few kilobases. The difference between the DNAs in lanes 4 and 8 is probably due to the fact that different bones were used in the two studies (femur vs. sternum and clavicula), but could also be the result of the described modifications of the extraction procedure.

3.2 Demonstration of the Presence of DNA

The fluorescent smears observed in Figure 2 could be due to substances other than DNA. In fact, the brown-colored contaminant found in the skin sample is fluorescent and shows a similar migration during agarose gel electrophoresis as degraded DNA. Therefore, it is important to consider methods that can be used to demonstrate the presence of DNA in samples more convincingly than ethidium bromide staining alone. Considering the small amount of material that is usually obtained from archaeological samples and the unknown nature of possible contaminants, most chemical and physical standard laboratory demonstrations of DNA are not adequate. Fortunately,

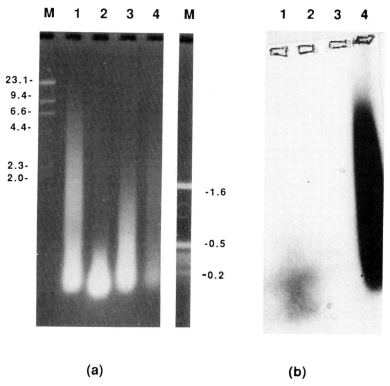

(a)   **(b)**

FIGURE 3. Electrophoresis of extracted DNA from skin tissue. (*a*) Lane 1: extracted material (same as Fig. 2, lane 8). Lanes 2–3: extracted DNA degraded by nonspecific and sequence-specific DNA endonucleases, respectively. Lane 4: extracted DNA challenged to DNase I and DNA polymerase in the presence of radioactively labeled dNTPs. (*b*) Same gel exposed to an X-ray film. Lane 4 demonstrates the successful incorporation of radioactivity. Numbers on either side of the gel indicate sizes (in kb) of DNA molecular weight markers. (Modified from Thuesen and Engberg 1989; reproduced with permission).

some of the highly specific tools of molecular biology, namely DNA-modifying enzymes, can be used in a simple way to unequivocally identify the nature of the fluorescent smears. In Figure 3a is shown a gel electrophoretic analysis of a mummy skin extract before (lane 1, comparable to Fig. 2, lane 8) and after (lanes 2–4) treatment with various enzymes. By observing the changes in electrophoretic mobility of the fluorescent material, we can conclude that the material is DNA: it is degraded by nonspecific (lane 2) and sequence-specific (restriction enzyme, lane 3) endonucleases, and it is resistant to RNase A (all samples were treated with RNase A prior to gel electrophoresis). The material in lane 4 was treated with DNase I and DNA polymerase I in the presence of a radioactively labeled dNTP mixture (nick-translation, Sambrook et al. 1989). The successful incorporation of radioactivity into the material as revealed by autoradiography of the gel (Fig. 3b) is conclusive evidence of the presence of template-competent DNA in the sample. An alternative approach to the demonstration of DNA in ancient bone was taken by Hummel and Herrmann (1991), who examined thin sections of bone under the microscope after staining with fuchsin.

3.3 Demonstration of Human DNA by Hybridization Analysis

A different, much more elaborate demonstration of DNA in extracts is hybridization analysis. A labeled probe, such as a radioactively labeled DNA fragment of known composition, is hybridized to the sample in solution or immobilized on a filter. Binding of the probe to the sample is evidence for complementarity of the two in terms of base-pairing ability and thus a measure of their relatedness. Hybridization analysis of DNA from archaeological samples is mainly used to demonstrate the presence of human DNA in the sample and, as the case may be, to reveal the species from which contaminating DNA originates.

In our analysis, we chose to probe the mummy material for the presence of two specific sequences that are found as highly repetitive DNA in the human genome. By choosing *Alu*-repeat and α-repeat sequences as probes, the hybridization analysis becomes far more sensitive than if unique sequences are used. The *Alu*-repeat is a ca. 300-bp repeat found in hundreds of thousands of copies throughout the human genome. The prototype α-repeat is 171-bp long and is similarly found in up to a million copies in the genome, preferentially in the centromeric regions of the chromosomes. The *Alu*- and α-repeat sequences together constitute 5–10% of the human genome. The enhanced sensitivity of the hybridization analysis was expected to be important because the Qilakitsoq mummies were covered with mold and much of the DNA could be expected to be of fungal origin.

Figure 4a shows the result of a hybridization of α-repeat probe to a Southern blot of a gel containing DNA from mummy skin and from human placenta used as control. A similar analysis of DNA from mummy bone is depicted in Figure 4b. In a series of control experiments (not shown, cf. Thuesen and Engberg 1990) it was found that these probes did not hybridize to DNA from *E. coli*, yeast, or mouse under the hybridization stringency conditions used. It was also demonstrated that radioactively labeled mummy DNA hybridized strongly to mummy DNA on filters showing the chemical integrity of this DNA, but weakly to human placenta DNA. This indicated that most of the mummy DNA was of nonhuman origin. In conclusion, the hybridization signals revealed in Figure 4a, b demonstrate the presence of human DNA in the mummy skin and bone samples.

The hybridization signals in experiments using the *Alu*- and α-repeat probes did not parallel the size distribution of ethidium bromide staining in gels, indicating that the DNA was a mixture of human DNA of relatively low

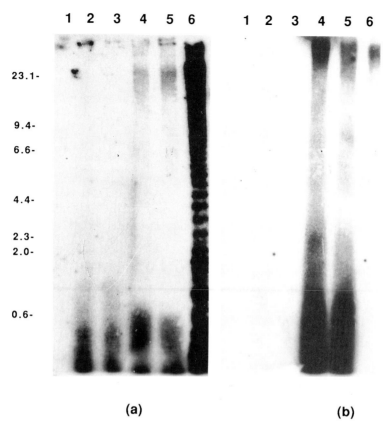

(a)   **(b)**

FIGURE 4. Hybridization analysis of DNA extracted from mummy bone and skin samples. All samples were digested with restriction enzyme *Alu*I prior to gel electrophoretic analysis. (*a*) Hybridization with the α-repeat probe. Lane 1: *E. coli* DNA. Lanes 2–3: DNA from bone. Lanes 4–5: DNA from skin. Lane 6: human placenta DNA. (*b*) Hybridization of the same samples to DNA from *Actinomycetes*. (From Thuesen and Engberg 1990; reproduced with permission.)

molecular weight and contaminating DNA of a higher molecular weight. Furthermore, the intensity of the hybridization signals of the autoradiograms shown in Figure 4 was far below what would be expected if the ethidium bromide stained DNA was mainly of human origin. By comparing the ethidium bromide staining and the hybridization signals from the mummy DNA with those from the human placenta DNA controls, we were able to estimate that only 0.002% of the DNA from mummy skin and 0.001% of the DNA from mummy bone was of human origin. It is possible that these estimates are low, because depurination and other chemical modifications that tend to weaken the ability of DNA to hybridize are expected to affect ancient human DNA to a larger extent than more recent contaminating DNA. The origin of the contaminating DNA was assessed by hybridizing the filters containing mummy DNA with radioactively labeled total DNA from different species. The mummy DNA did not hybridize to *E. coli* DNA, but at least a fraction of the contaminating DNA in skin samples hybridized to DNA from *Actinomycetes* (Fig. 4c). *Actinomycetes*, *Penicillium*, *Aspergillus*, and *Botrytis* have been positively identified in isolates from the mummies (Svejgaard et al. 1989).

The gel shown in Figure 2 was blotted onto a hybridization membrane and hybridized with an *Alu* probe (results not shown). Compared to the experiments described above, in this case the hybridization pattern paralleled the ethidium bromide staining pattern to a much higher degree. This result indicates that not only the extracted DNA as such, but also the DNA of human origin, i.e., the mummy DNA, was more intact in this experiment. The "hydroxyapatite protocol" is therefore expected to yield DNA of a better quality than our previous DNA extraction protocols.

3.4 Amplification of Specific DNA Segments by PCR

The polymerase chain reaction (PCR) is a particular useful method in studies of aDNA, because of the low amount of DNA in the samples and because of the various types of modifications of the aDNA that tend to lower considerably the efficiency of other methods, such as DNA cloning. We have used PCR in sex determination and in determining the ethnic origin of the Qilakitsoq mummy.

For sex determination we used two sets of oligonucleotide primers complementary to X-chromosome- and Y-chromosome-specific α-repeat sequences, respectively (Waye and Willard 1985; Witt and Erickson 1989). Thus the criterium for sex determination is the presence or absence of a Y-chromosome-specific PCR product. A gel electrophoretic analysis of the PCR products from amplification of mummy DNA is shown in Figure 5. An amplification product of 130-bp resulting from amplification using the X-chromosome-specific primers is seen in lane 1, whereas the 170-bp product expected from amplification using the Y-chromosome-specific primers is absent in lane 2. This is not due to an inability to amplify segments of this size from the mummy DNA: even larger segments have been amplified from this sample (see below). The result, demonstrating that mummy I/5 was a female, is in accordance with the anthropological observations.

In order to assess the ethnic origin of the mummy, we carried out PCR amplifications from mitochondrial DNA (mtDNA) and subsequently sequenced the PCR products. The mitochondrial genome is characterized by a much higher effective mutation rate than the nuclear genome, by maternal inheritance, and by apparent haploidy. The copy number of mtDNA is high (an average somatic cell contains thousands of mtDNA molecules), and it is possible that organellar DNA is better preserved than nuclear DNA in aDNA samples. Altogether, these facts make mtDNA a highly suitable object for studies of ethnic origin in aDNA samples.

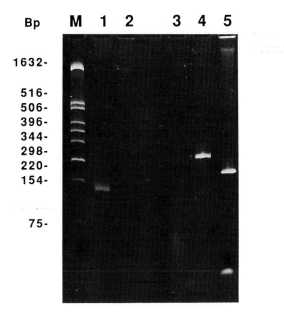

FIGURE 5. Gel electrophoretic analysis of PCR products resulting from amplification of a mummy bone extract. The PCR reactions were performed using specific oligonucleotide primers and reagents from a GeneAmp kit (Perkin Elmer Cetus). Temperature cycling parameters were: 1′ at 94°C, 2′ at 50°C, 2′ at 72°C, and 35 cycles. The reactions were "hot-started" by the addition of 5 U of *Taq* DNA polymerase at 72°C. The cycling was done using a Bioexcellence thermal incubator. 10/200 μl of the extract (the same as analyzed in Fig. 2, lane 8) was added to all reactions, and 10/50 μl of the amplification reaction was analyzed by electrophoresis on a 5% polyacrylamide gel. Lane 1: X-chromosome-specific α-repeat primers. Lane 2: Y-chromosome-specific α-repeat primers. Lane 3: mtDNA primers L15997 and H16401. Lane 4: mtDNA primers L16099 and H16255. Lane 5: amplification of human control DNA (XY) using Y-chromosome-specific α-repeat primers. M: molecular weight marker pBR322 × *Hin*fI. Note that the size of the DNA fragments is the segment amplified plus the size of the primers. The low molecular weight fragment in lane 5 is probably an artifact known as a "primer-dimer."

Most of the variability in mtDNA is found in the noncoding control region. We have chosen to focus on the so-called D-loop region within the control region because more than half of the variability is concentrated in this short segment (Orrego and King 1990). The primers used in PCR amplification of mtDNA are identified by a letter designating the strand of mtDNA (*Light* or *Heavy*) and a number corresponding to the reference sequence coordinate of the base at the 3′-end of the primer. The reference sequence is the first complete human mtDNA sequence published (Anderson et al. 1981). We were able to amplify mtDNA segments of more than 150 bp (Fig. 5, lane 4) but not segments as long as 400 bp (Fig. 5, lane 3) from the mummy DNA

Bp   M 1 2 3 4 5

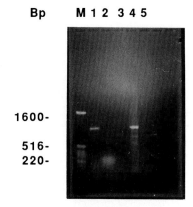

1600-

516-
220-

FIGURE 6. Amplification of a mtDNA- segment from two different bone extracts. The experimental details are the same as in Figure 5 except that the PCR products were analyzed on a 1% agarose gel. The mtDNA primers used were L16209 and H00580. Lane 1: extract prepared by the "hydroxyapatite" protocol (Section 3.1). Lane 2: extract prepared by the "high EDTA" protocol. Lane 3: mock extract from the "hydroxyapatite" protocol. Lane 4: human control DNA. Lane 5: no DNA. M: pBR322 × HinfI.

extracted in our earlier study (Thuesen and Engberg 1990). The DNA preparation did not contain substantial amounts of *Taq* DNA polymerase inhibitors since amplification of a well-characterized DNA template added to the sample was successful. Pre-repair of the DNA was not attempted. A number of standard PCR parameters were systematically varied, without success. Figure 6 shows a similar experiment using recently extracted DNA as described in Section 3.1. The primers in this case were L16209 and H00580, and the expected 980-bp PCR product was found when DNA extracted according to the third extraction protocol was used (lane 1, the "hydroxyapatite protocol") but not when DNA extracted by the first protocol was used (lane 2, the "high EDTA" protocol). No PCR product was formed in a reaction involving mock extract from the third protocol (lane 3) or in a PCR reaction to which no extract was added (lane 5). This result implies that the DNA eluted from bone powder under standard conditions for elution of double-stranded DNA from hydroxyapatite is of a better quality in terms of template competence than the DNA extracted according to our previously published protocol.

3.5 Sequence Characterization of PCR Products

The preferred strategy for sequencing of PCR products is to sequence the products directly rather than after cloning. Replication errors due to modifications of the aDNA template and errors introduced because of the high inherent error rate of the *Taq* DNA polymerase will affect individual molecules and their descendants and will be canceled out when the whole population of molecules is sequenced. In contrast, if the PCR product is cloned and the sequences are derived from an individual clone, all molecules in the sequencing reaction will be affected and their sequence is likely to be erroneous. In this case, it is imperative to sequence DNA from several independent clones.

In our early studies, we were unable to obtain complete, clean sequences from direct sequencing of PCR products. Instead, the PCR products depicted in Figure 5 were cloned and DNAs from 5 to 10 individual clones were sequenced. For cloning, PCR products were purified on Millipore Ultrafree-MC filters. The DNA was phosphorylated using polynucleotide kinase and blunt-ended using the 3′-5′ exonuclease and the 5′-3′ polymerase activities of Klenow polymerase. The PCR products were cloned into the *Sma*I site of a Bluescript vector (Stratagene) by conventional methods (Sambrook et al. 1989). Finally, DNA was isolated from positive clones and sequenced.

Using this strategy, we have sequenced the X-chromosome α-repeat and the mitochondrial D-loop segment from mummy DNA and a Caucasian control. The α-repeat sequences differed in 1 of 90 positions (not shown). The D-loop was strikingly similar to the Caucasian reference sequence, differing in only 1 of 155 positions (Fig. 7)—a position furthermore known to be polymorphic in Caucasians (Orrego and King 1990). This observation led us to suspect that the mummy sequence was derived from contaminating Caucasian DNA. However, nothing was known at that time about Eskimo D-loop sequences. We therefore sequenced the relevant D-loop segment from contemporary Eskimos. Samples were obtained from five persons from west Greenland and five from east Greenland, all with a documented Eskimo maternal background. PCR amplification was made from DNA extracted from blood samples or directly from 2 μl of whole blood (Mercier et al. 1988). The PCR products were purified as described above and both strands sequenced directly by cycle sequencing using a kit (BRL). The sequences are displayed in Figure 7. Six different sequence types were found. All of these resembled the Caucasian reference sequence in the sense that no major polymorphism (e.g., duplication or deletion) was found. The sample is too small to allow a meaningful calculation of nucleotide frequencies at the polymorphic positions in order to characterize the Eskimo population. However, the variability seems to be of the same order as that reported by Stoneking et al. (1991) who sequenced the entire control region from 52 individuals. Their study included Caucasians, Africans, Asians, and others. The polymorphisms in the Eskimos at pos. 16111, 16114, 16192, and 16223 have also been found in Caucasians (Orrego and King 1990), whereas pos. 16100, 16241, and 16250 were found to be invariable by Stoneking et al. (1991) and Orrego and King (1990).

In conclusion, we are not yet in a position to characterize a sequence such as the mummy D-loop sequence as being of Eskimo type. The similarity to the Caucasian reference sequence is not disturbing in view of the sequences

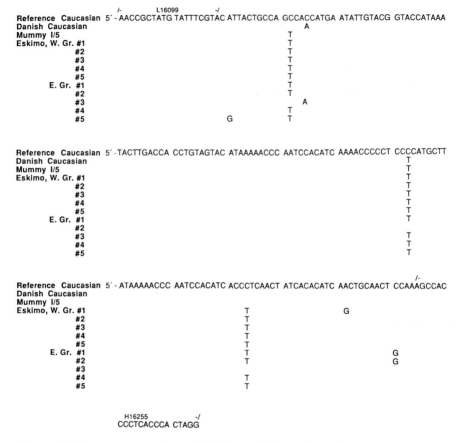

FIGURE 7. Sequence comparison of Eskimo and Caucasian D-loop sequences. The segment of mtDNA D-loop sequence between primers L16099 and H16255 is shown for the Caucasian reference sequence (Anderson et al. 1981), a Danish Caucasian, mummy I/5, and contemporary Eskimos from west (W) and east (E) Greenland. Only the sequence differences relative to the reference sequence are displayed.

obtained from contemporary Eskimos. Although we are unable to prove that the mummy sequence is genuine, we are confident that this is the case. First, our hybridization results (Figs. 3 and 4) are unlikely to have been caused by contamination. Second, our inability to amplify Y-chromosome-specific DNA segments, combined with the successful amplification of larger mtDNA segments, suggests that the DNA was of female origin. Since the DNA was handled by males exclusively, this observation strengthens the case for authenticity of the mummy DNA.

4. Discussion

The field of aDNA studies is still in its infancy. Nevertheless, it is important now to move from mere fascination with the existence of DNA in ancient samples to systematic methodological studies and, more importantly, to the application of aDNA analysis to specific problems in anthropology and other sciences. One example of such a problem, taken from the Arctic, is the colonization history of Greenland. Greenland was first populated by the Paleoeskimos (the Dorset culture). About a thousand years ago, these were replaced by Eskimos of the Thule culture in the north and by the Norse populating the south. Later in history, the Thule Eskimos expanded and the Norse disappeared. From then on colonies were formed by Europeans. Some of the questions that remain unsolved are: What happened to the Dorset Eskimos? What migration routes did the Thule Eskimos follow when they settled in Greenland? How are we to explain the apparent anthropological differences in the modern population of Greenland? Did the Eskimo and the Norse population mix? It would be of great interest if studies of DNA in finds from Greenland, combined with population studies of contemporary Eskimos, could answer some of these questions.

The first step toward success in aDNA studies is good field work. It is important that the advantages offered by frozen samples in the Arctic not be lost to post-recovery deterioration and contamination. According to our own experience with excavations in Syria, coordination of traditional work and DNA work at an archaeological site causes only minor problems. However, finds in the Arctic and Alpine areas are often sporadic (e.g., the Qilakitsoq mummies and Ötze, the Glacier Man). Thus general awareness of a set of simple guidelines would be most helpful in order to avoid contamination and to keep the material at the favorable low temperatures.

The second important step in aDNA work is of course the protocols for recovery of DNA of good quality. Relatively few studies of recovery of DNA from frozen or temporarily frozen samples have been reported. In addition to our own work, these are mainly the sex determination by PCR amplification of bone extracts applied to Spitsbergen burials (Hummel and Herrmann 1991) and the work by Higuchi on mammoths from the Siberian tundra. The latter work has been taken to the sequencing level and 350 bp of mtDNA sequence has been obtained. It differs at four or five positions from the Indian elephant sequence, and by a similar amount from the African elephant, confirming the three-way split (Cherfas 1991).

In this chapter, we have proposed a new idea for extraction of DNA from bones based on elution conditions for DNA bound to hydroxyapatite. Our results are based on extraction from a single bone sample. Although the proper controls were included, the method needs further testing before it can be evaluated. One area that requires further investigation is the effect, in terms of DNA binding properties, of the conversion of hydroxyapatite to brushit and other calcium phosphates that takes place during diagenesis (Herrmann et al. 1990).

aDNA is degraded into small fragments and contains an abundance of lesions such as baseless sites, deaminated bases, oxidized bases, and cross-links (Pääbo 1989). It is highly desirable that methods be developed to evaluate DNA lesions prior to analysis. The application of such methods would be helpful in conducting pre-repair of the DNA (see, e.g., Rogan and Salvo 1990) and in planning PCR strategies, and it could hint at the authenticity of the resulting sequence. A standard curve for monitoring apurinic sites can easily be derived from depurination of DNA by 0.1 N HCl, followed by strand cleavage in alkali or by RNase A. Oxidative DNA damage can possibly be assayed by the application of DNA repair enzymes such as AP endonuclease and DNA glycosylases (Wallace 1988; Pääbo 1989). Monitoring the results of template switches ("jumping PCR"; cf. Oste, Chapter IV.2) in PCR amplification reactions as a measure of DNA integrity has been suggested by Pääbo et al. (1990).

The recovery of high-quality DNA is the bottleneck in aDNA studies. If high-quality DNA is obtained, many of the advanced methods of molecular biology can be let loose on the DNA. With respect to the problems of the colonization of Greenland described above, one important question is whether an ethnic criteria can be applied to a single specimen. If high-quality DNA is available, many different approaches can be taken to this problem. Medical records in Greenland are comprehensive, and this has stimulated research in disease-associated polymorphisms. Characteristic allele frequencies in the HLA-system have been described, including one example of an allele found in Eskimos but so far not in Scandinavians (L. Lamm, pers. comm.). One complicating factor is that the proportion of Caucasian genes in Greenland today is high. In western Greenland it was estimated to be least 25–30% in the 1970s based on HLA studies (Kissmeyer-Nielsen et al. 1971). This contribution of Caucasian genes is predominantly paternal and does not affect the maternally inherited mtDNA to the same extent. For a number of other reasons outlined in Section 3.4, we prefer to focus our interest in criteria for ethnic origin on mtDNA. One would hope to find a major, diagnostic polymorphism such as the 9-bp direct repeat, which occurs in only one copy in many Asians as well as in some Native Americans (Pääbo et al. 1988; Wrishnik et al. 1987). In the absence of such diagnostic, the SSO-typing system of Stoneking et al. (1991) can be applied. The amount of heterozygosity in this system is greater than that detected in highly polymorphic loci in nuclear DNA, such as the HLA and VNTR systems. Although the system is based on frequencies of polymorphic types rather than diagnostic sequence differences, it can probably be used to give answers with high probability values because the populations that need to be considered in this case (Eskimos and Caucasians) are few and only distantly related.

Acknowledgments. The sample of mummified bone used in this study was supplied by Dr. J.P. Hart Hansen. The DNA and blood samples from modern Eskimos were generously supplied by Drs. Lars Lamm, Peter Steen Hansen, and Søren Nørby. We are grateful to Drs. Margaret and Jørgen Christoffersen, Pia Bennike, and Niels Lynnerup for helpful discussions and to Mr. Franz Frenzel and Ms. Else Uhrenfeldt for technical assistance and typing of the manuscript, respectively.

References

Anderson S, Bankier AT, Barrell BG, de Brujn MHL, Coulson AR, Drouin J, Eperon IC, Nierlich DP, Roe BA, Sanger F, Schreier PH, Smith AJH, Staden R, Young IG (1981) Sequence and organization of the human mitochondrial genome. Nature 290:457–465

Artamanov ML (1965) Frozen tombs of the Scythians. Sci Am 212:161–109

Aufderheide A (1981) Soft tissue paleopathology: an emerging subspeciality. Hum Pathol 12:865–867

Cherfas J (1991) Ancient DNA: still busy after death. Science 253:1354–1356

Coghlan A (1992) Ötze: the man who came in from the cold. New Sci 11 (January): 17–18

Hart Hansen JP (1989) The mummies from Qilakitsoq: paleopathological aspects. Meddr Grønland, Man & Soc 12:69–82

Hart Hansen JP, Gulløv HC (eds) (1989) Meddr Grønland, Man & Soc 12

Herrmann B, Grupe G, Hummel S, Piepenbrink H, Schutkowski H (1990) *Prähistorische Anthropologie*. Berlin: Springer-Verlag

Horne P (1980) The death of Charles Francis Hall, Arctic explorer (abstr.). Annual Meeting of the Paleopathological Association, p. NF 2, Niagara Falls, New York

Hummel S, Herrmann B (1991) Y-chromosome-specific DNA amplified in ancient human bone. Naturwissenschaften 78:266–267

Kissmeyer-Nielsen F, Andersen H, Hauge M (1971) HLA types in Danish Eskimos from Greenland. Tissue Antigens 1:74–80

Mercier B, Gaucher C, Feugeas O, Mazurier C (1988) Direct PCR from whole blood, without DNA extraction. Nucl Acids Res 18:5908

Micozzi MS (1986) Experimental study of postmortem change under field conditions: effects of freezing, thawing and mechanical injury. J Forensic Sci 31:953–961

Micozzi MS (1991) Postmortem change in human and animal remains: a systematic approach. Springfield, Ill.: Thomas Books

Newell RR (1984) The archaeological, human biological, and comparative conexts of a catastrophically-terminated Kataliguaq house at Utqiagvik, Alaska (BAR-2). Arctic Anthropol 21:5–51

Orrego C, King MC (1990) Determination of familial relationships. In: Innis MA, Gelfand DH, Sninsky JJ, White TJ (eds) *PCR Protocols: A Guide to Methods and Applications*. San Diego: Academic Press, pp. 416–426

Pääbo S (1986) Molecular genetic investigations of ancient human remains. Cold Spring Harbor Symp Quant Biol 51:441–446

Pääbo S (1989) Ancient DNA: extraction, characterization, molecular cloning, and enzymatic amplification. Proc Natl Acad Sci USA 86:1939–1943

Pääbo S (1990) Amplifying ancient DNA. In: Innis MA, Gelfand DH, Sninsky JJ, White TJ (eds) *PCR Protocols: A Guide to Methods and Applications*. San Diego: Academic Press, pp. 159–166

Pääbo S, Gifford JA, Wilson AC (1988) Mitochondrial DNA sequences from a 7000-year old brain. Nucl Acids Res 16:9775–9787

Pääbo S, Higuchi RG, Wilson AC (1989) Ancient DNA and the polymerase chain reaction. J Biol Chem 264:9709–9712

Pääbo S, Irwin DM, Wilson AC (1990) DNA damage promotes jumping between templates during enzymatic amplification. J Biol Chem 265:4718–4721

Prager EM, Wilson AC, Lowenstein JM, Sarich VM (1980) Mammoth albumin. Science 209:287–289

Rogan PK, Salvo JJ (1990) Molecular genetics of Pre-Columbian South American mummies. UCLA Symp Mol Cell Biol 122:223–234

Rudenko SI (1970) *Frozen Tombs of Siberia*. Edited and translated by MV Thompson. Berkeley, Cal.: University of California Press

Sambrook J, Fritsch EF, Maniatis T (1989) *Molecular Cloning: A Laboratory Manual*. 2nd ed. Cold Spring Harbor, N.Y.: CSHL Press

Sillen A (1989) Diagenesis of the inorganic phase of cortical bone. In: Price TD (ed) *The Chemistry of Prehistoric Human Bone*. Cambridge, U.K.: Cambridge University Press, pp. 211–229

Stoneking M, Hedgecock D, Higuchi RG, Vigilant L, Erlich HA (1991) Population variation of human mtDNA control region sequences detected by enzymatic amplification and sequence-specific oligonucleotide probes. Am J Hum Genet 48:370–382

Svejgaard E, Stenderup A, Møller G (1989) Isolation and eradication of fungi contaminating the mummified corpses from Qilakitsoq. Meddr Grønland, Man & Soc 12:131–133

Thuesen I, Engberg J (1990) Recovery and analysis of human genetic material from mummified tissue and bone. J Arch Sci 17:679–689

Wallace SS (1988) AP endonuclease and DNA glycosylases that recognize oxidative DNA damage. Environ Mol Mutagen 12:431–477

Waye JS, Willard HF (1985) Chromosome-specific alpha satellite DNA: nucleotide sequence analysis of the 2.0 kilobasepair repeat from the human X-chromosome. Nucl Acids Res 13:2731–2743

Witt M, Erickson RP (1989) A rapid method for detection of Y-chromosomal DNA from dried blood specimens by the polymerase chain reaction. Hum Genet 82:271–274

Wrischnik LA, Higuchi RG, Stoneking M, Erlich HA, Arnheim N, Wilson AC (1987) Length mutations in human mitochondrial DNA: direct sequencing of enzymatically amplified DNA. Nucl Acids Res 15:529–542

Zimmerman MR, Smith GS (1975) A probable case of accidental inhumation 1600 years ago. Bull NY Acad Med 51:828–837

Zimmerman MR, Yearman GW, Sprinz H (1971) Examination of an Aleutian mummy. Bull NY Acad Med 47:80–103

Zimmerman MR, Trinkaus E, LeMay M (1981) The paleopathology of an Aleutian mummy. Arch Path Lab med 105:638–641

Dried Samples: Body Fluids
9 DNA Typing of Biological Evidence Material

GEORGE F. SENSABAUGH

1. Introduction

It has been amply demonstrated that DNA can persist in dried biological materials for many years, in some cases decades and centuries. The secret to this persistence is the absence of bulk water in such materials, for the most damaging mechanisms of DNA deterioration require water either as a reactant or as a medium. In the absence of water, the rates of microbial growth, enzymatic degradation, spontaneous hydrolysis, and other forms of chemical damage are greatly slowed.

This chapter describes the extraction of DNA from dried residues of body fluids. Although these procedures have been developed primarily in the context of the forensic analysis of biological evidence (i.e., blood stains, semen traces, etc.), the principles are generally applicable to any form of dried biological material. Biological evidence, in fact, may be looked upon as a precursor to more aged forms of biological samples. Both kinds of materials have histories which include contact with surfaces (metal, stone, fabrics, soil, etc.), exposure to the lighting, temperature, and humidity conditions of the surrounding environment, and possibly exposure to a wide range of chemical and biological contaminants (microbes, dust, insects, etc.).

The objectives in the analysis of biological evidence and ancient biological samples are also consonant. Forensic testing may be undertaken to determine species, sex, and/or individual origin; the same objectives apply to ancient samples. Testing at the DNA level employs the same basic approach, the detection of a relevant informative sequence; the target sequence can be detected either by a specific probe or by specific amplification. The best-studied sequence for species testing is the cytochrome b gene on mitochondrial DNA; the data base of species-specific sequences is rapidly expanding (see, e.g., Irwin et al. 1991). For sex testing, a number of X- and Y-chromosome-associated sequences have been identified (see, e.g., Witt and Erickson 1989; Aasen and Medrano 1990). For identity testing, a wide range of options exist; virtually the entire variability of the genome is accessible.

2. Basic Considerations

2.1 DNA Packaging in Cells

Most of the DNA in animal cells is contained in the cell nucleus. Nuclear DNA in somatic cells is complexed with specific nuclear proteins; these complexes are referred to as chromatin (Igo-Kemenes et al. 1982). The major DNA binding proteins in chromatin are the histones, a family of lysine- and arginine-rich proteins. The nuclear DNA of sperm is packaged differently; it is complexed with another family of proteins, the protamines (Ward and Coffey 1991). The DNA-protamine complexes in sperm are more condensed than nuclear chromatin and are disulfide crosslinked. The practical consequence of this difference in packaging is that nuclear DNA can be released from chromatin by proteolytic digestion in the presence of a detergent, whereas the release of sperm DNA requires in addition the presence of a disulfide-reducing agent. This allows differential extraction of somatic and sperm DNA in samples containing mixtures of the two cell types.

Animal cells also contain a mitochondrial genome; this genome is encoded on a piece of circular DNA (mtDNA) containing about 16,600 bp. The distribution of mitochondria in somatic cells and sperm differs, again with practical consequences. Somatic cells harbor 500–1,000 mitochondria per cell, each containing a copy of the mtDNA. Under normal whole cell extraction conditions, the nuclear and mitochondrial DNA of somatic cells co-extract. Sperm contain fewer than 100 mitochondria, all of which are located in the sperm tail. Sperm mtDNA can be extracted under normal somatic cell extraction conditions whereas, as noted above, sperm nuclear DNA cannot. Accordingly, the nuclear and mitochondrial DNA of sperm can also be differentially extracted.

2.2 DNA Preservation in Biological Evidence

The conditions to which a biological evidence sample is exposed prior to its arrival at the laboratory may affect the quantity and quality of recoverable DNA. The DNA may be partially or extensively degraded, may have undergone some chemical modification to the base and/or sugar units, and may have undergone some degree of crosslinking.

The extent of DNA degradation, i.e., strand breakage, in biological evidence samples reflects the history of the sample. Degradation is usually mediated by the action of endogenous nucleases prior to the drying of the sample. Microbial contaminants can also provide nucleases leading to degradation. DNA degradation occurs via hydrolysis; in dry samples, the rate of degradation is slow. Accordingly, spontaneous degradation is not believed to contribute significantly to degradation patterns in dry evidence samples.

The factors contributing to base modification and DNA–DNA or DNA–protein crosslinking have been described (Friedberg 1985; Sensabaugh and von Beroldingen 1991). Examples of nucleotide modification include pyrimi-

dine dimer formation resulting from exposure to UV radiation, base and/or sugar modifications resulting from oxidation reactions, and base deamidation. The consequence of nucleotide modification for subsequent analyses depends on the nature and extent of the modification. Tests with pyrimidine dimer formation, for example, indicate that the damage must be near saturation levels to have an effect (Buoncristiani et al. 1990). DNA crosslinking reduces extraction efficiency and, in extreme cases, renders the DNA unextractable.

Microbial contamination of evidence samples may occur by virtue of the site of origin (e.g., in samples collected from the mouth or vagina) or as a consequence of microbial growth on the sample. The latter occurs only when conditions are favorable, i.e., when the sample remains moist for an extended period of time. Microbial contamination has two consequences for DNA analysis: the contribution of degradative enzymes and the contribution of extraneous DNA. In some cases, evidence samples may contain much more DNA from contaminating microbes than from the original source.

Once a sample enters the laboratory, there is no excuse for additional deterioration to occur. As pointed out above, most deterioration reactions require water; accordingly, storage in the dry state is preferred. Also, deterioration reactions occur only at a negligible rate at freezer temperatures; thus freezer storage is recommended. Since samples removed from the freezer will collect condensation moisture, it is advisable to place such samples in a desiccator until room temperature is reached.

2.3 Assessment of DNA Quantity and Quality

The quantity and quality of DNA in an evidence sample determines the subsequent course of analysis. Restriction fragment length polymorphism (RFLP) analysis requires samples containing at least 10,000 copies of relatively intact DNA; this amount corresponds to about 50 ng of human genomic DNA. Analytical methods based on the polymerase chain reaction (PCR), on the other hand, can be used on samples containing as few as 10–20 copies of DNA template (ca. 50–100 pg human genomic DNA) and on samples containing significantly degraded DNA.

The quantity and extent of degradation of DNA in a sample is readily assessed by electrophoresis of the extracted DNA on a minigel (see e.g., Sambrook et al. 1989). With ethidium bromide staining, as little as 3 ng of relatively intact DNA can be detected; detection sensitivity is lower for degraded DNA. For samples believed to contain significant amounts of contaminating (microbial) DNA, the quantity of DNA from the species of interest can be determined by probing with a species-specific probe in a dot blot format (Weye et al. 1989); the use of probes against repetitive sequences (e.g., alpha satellite or Alu sequences) extends sensitivity to sub-nanogram levels. Electrophoretic characterization followed by Southern blotting and probing with a species-specific probe allows assessment of the state of degradation of the target DNA (unpul. observations).

2.4 Interferences

Some biological evidence samples yield DNA that on first pass is refractory to analysis. The most common interference is inhibition of the restriction enzymes used in RFLP analysis or the polymerases used in PCR. Interfering enzyme inhibitors have been found to co-extract with the DNA from old blood stains, hair, and samples containing certain dyes; the inhibitors are believed to be the pigments in these samples, i.e., free heme from blood stains, melanin pigments from hair, and dyestuffs such as indigo. As a general rule, if the DNA from a sample contains a co-extracted pigment, interference with restriction and amplification can be expected. Cations bound to DNA can also interfere with restriction and amplification.

In some instances, inhibitors can be removed by repeated DNA isolation or by modification of the extraction procedure. An example of the latter is the use of the chelating resin Chelex for the preparation of DNA for PCR (Walsh et al. 1991). In other cases, the interference can be overcome by modification of the analytical procedure. For example, the interference of some PCR polymerase inhibitors can be overcome by the addition of serum albumin to the amplification mix (Pääbo et al. 1988); albumin complexes many compounds, including heme, dyes, and fatty acids. Another approach is to use excess polymerase in the amplification reaction.

2.5 Sample Handling

The amount of DNA in biological evidence samples is often small, and extraction procedures must be modified accordingly. Standard large-scale DNA isolation generally employs a phenol/chloroform extraction followed by ethanol precipitation (Sambrook et al. 1989). On a small scale, however, DNA is often lost at the precipitation step. An alternative is the use of disposable spin ultrafiltration tubes such as the Centricon 100 tube; the DNA fraction can be washed and concentrated into a small volume in a single tube (Higuchi et al. 1988). This procedure is recommended for samples containing less than 500 ng DNA. The Chelex isolation procedure mentioned above also provides a one-step DNA isolation procedure for PCR analysis; it cannot be used for RFLP analysis since it yields single-stranded DNA product.

Dried stains on nonporous surfaces are best scraped off and the flakes used for analysis. Stains on porous materials such as fabrics may be soaked in a small volume of water or TE buffer to elute the stain; DNA is extracted from the aqueous suspension after removal of the fabric. However, because cells often adhere to such supports, it is usually desirable to extract the stained material directly.

Forensic laboratories must be particularly cautious to avoid sample handling errors, i.e., sample mix-ups and sample contamination. This can be addressed by allowing only one sample to be prepared at a time; the risk of a handling error is reduced at the price of efficiency. With PCR analysis, there is also a concern regarding sample contamination by PCR amplification products ("carry-over" contamination); the risk is that the products of one analysis could "seed" subsequent amplifications in other samples. This problem can be addressed by careful sample handling and by directing the flow of analysis in a single direction such that the products of analysis have no opportunity to come into contact with samples waiting to be analyzed (Reynolds et al. 1991).

3. DNA Isolation

The procedures described below are derived from Gill et al. (1985), Higuchi et al. (1988), and Walsh et al. (1991).

3.1 General Procedure for Blood, Hair, and Tissues

3.1.1 Solutions and Materials

Stock Solutions:
  1 M Tris-HCl, pH 7.5
  10% SDS
  0.5 M Na$_2$EDTA, pH 8.0
  5.0 M NaCl
  3 M Na Acetate (NaAc)
  1 M DTT in 10 mM NaAc, pH 5.2
  Proteinase K, 10 mg/ml in H$_2$O
  Phenol/Chloroform/extraction solvent
  Absolute ethanol (EtOH)
  TE buffer, pH 7.5

Digest Buffer (made from stock solutions):

	For 100 ml	For 0.5 ml
10 mM Tris-HCl	1 ml	5 µl
10 mM EDTA	2 ml	10 µl
50 mM NaCl	1 ml	5 µl
2% SDS	20 ml	100 µl
sterile deionized H$_2$O (dH$_2$O)	76 ml	380 µl

Microconcentrators: Centricon-100 microconcentrators (Amicon) or any other microconcentrators with appropriate filtration limits and nonadsorbent filters may be used.

3.1.2 Procedure for Blood and Blood Stains

1. Suspend 0.1 ml whole blood, a white cell fraction, or a blood stain in 0.5 ml digest buffer. Add 15 µl proteinase K solution. Incubate for at least 1 hr at 56°C. Old blood stain samples are sometimes best incubated

overnight at 37°C followed by addition of a second aliquot of proteinase K and a second incubation. Remove blood stain substrate.

2. Add 0.5 ml buffered phenol/chloroform. Mix by inverting ca. 20 times. Centrifuge 1–2 min in a microfuge (10–12,000 rpm).
3. Remove aqueous upper layer and transfer to new tube. Repeat extraction until interface is clear, usually after 2–3 extractions. To maximize yield, material appearing at the interface should be transferred with the aqueous phase for several extractions.
4. Remove residual chloroform by extracting with an equal volume of n-butanol. This is essential if Centricon microconcentrators are to be used; chloroform destroys the filtration membrane.
5. Remove lower layer and isolate/concentrate DNA by Centricon-100 dialysis filtration or EtOH precipitation as described below. Use of a microconcentrator is recommended if the sample contains less than 500 ng DNA.

3.1.3 Procedure for Hair and Tissues

1. Wash hairs with deionized water, then with EtOH; dry. For mounted hairs, freeze slide, remove cover slip with a scalpel, and wash away mounting medium with xylene. Wash hair with EtOH, then deionized water, then EtOH.
2. Digest hair for several hours at 37°–56°C with digest buffer, 20 µl DTT, and 15 µl proteinase K. If the hair does not dissolve after this initial incubation, add additional aliquots of DTT and proteinase K and incubate an additional several hours or overnight until hair is completely dissolved. Centrifuge to collect or remove pigment.
3. Extract with phenol/chloroform, then n-butanol as described above, and concentrate DNA fraction.

Note: Tissue samples (boiled or well minced), bone, etc. can be treated similarly to hair.

3.1.4 Concentration Steps

1. *Ethanol Precipitation.* Transfer aqueous DNA extract to microfuge tube or other appropriate centrifuge tube. Add 1/10 volume 3 M NaAc, then add 2.5 volumes cold 100% EtOH. Store at − 20°C for 1 hr or overnight. Collect DNA by centrifugation, 10–30 min, in a microfuge (10–12,000 rpm).
2. *Centricon 100 Ultrafiltration.* Add 1.5 ml TE buffer to Centricon tube. Add butanol-extracted DNA solution. Centrifuge using a fixed-angle rotor for 20 min at 4–5,000 rpm. Add 2 ml TE buffer to concentrated DNA (now at ca. 40 µl). Repeat centrifugation. Wash DNA 2–3 times.

5.2 Differential DNA Extraction from Sperm and Epithelial Cells

1. Suspend vaginal swab or semen stain in 1 ml dH$_2$O in a 1.5-ml microfuge tube for 30 min at room temperature. Agitate the stained material with a sterile probe to loosen cellular debris and squeeze out the stain against the wall of the tube; set the stained material aside. Centrifuge and view a small portion of the debris microscopically for the presence of sperm.
2. Suspend the pellet containing sperm and cellular debris in 0.5 ml digest buffer and add back the stain material. Add 15 μl proteinase K and digest at 56°C *for no more than 2 hr.* Centrifuge the digest for 5 min at 10,000–12,000 rpm. The supernatant contains the epithelial cell DNA; this is removed for analysis.
3. Wash pellet 2–3 times with digest buffer by centrifugation; discard the supernatants. Wash pellet once with water and remove all but about 50 μl. Suspend the pellet in this reduced volume and remove a small aliquot of sediment (5–10%) for microscopic exam to verify digestion of epithelial cells and recovery of sperm.
4. To the pellet suspension, add 0.5 ml digest buffer. Add 20 μl of 1 M DTT, 15 μl proteinase K, and incubate for at least 1 hr at 56°C. The completeness of the digestion can be verified by spinning down any remaining cellular debris and examining the debris microscopically. The digest supernatant contains the sperm DNA.
5. The two DNA fractions are extracted with phenol/chloroform and subsequently processed as in the general procedure described above.

Note: Prolonged first digestion will cause sperm to swell and lead to loss of sperm DNA. This sample preparation step requires some skill and good judgment. The wash steps are critical, particularly when the ratio of sperm to epithelial cells is small.

3.3 Chelex Extraction for PCR Analysis

Chelex (Bio-Rad Laboratories) is a chelating resin with a high affinity for polyvalent metal ions. It comes in the sodium form, 100–200 mesh. The resin must be pipetted while in suspension; this can be achieved by pipetting with a large-bore pipet tip from a stirring suspension. The stock solutions are 5% and 20% suspensions of Chelex (w/v) in sterile distilled water.

The basic procedure is to incubate the sample at 100°C in the presence of 5% Chelex, centrifuge down the resin and cellular residue, and transfer an aliquot of the supernatant solution to the PCR amplification mix. Chelex supernatant solutions can be stored for several weeks with refrigeration or freezing; because the supernatant solutions are alkaline, prolonged storage is not advisable.

For blood stains on fabrics, a small cutting is soaked in 1 ml dH$_2$O for 15–30 min, then centrifuged. The supernatant is removed, leaving the stain material and pellet in the tube. Equal volumes of dH_2O and 10% Chelex suspension are added, the mixture vortexed, incubated at 56°C for 15–13 min, vortexed again, and then placed in a boiling water bath for 8 min. The mixture is vortexed again, centrifuged, and an aliquot of the supernatant solution is removed for amplification.

In differential extractions, the epithelial cell fraction (step 2 above) is mixed 3:1 with 20% Chelex stock solution, incubated at 100°C for 8 min, and then processed as described above. To the 50 µl sperm suspension (step 3), add 150 µl 20% Chelex stock suspension, 2 µl proteinase K, and 7 µl DTT stock solution; the mixture is incubated at 37°C for 30–60 min, vortexed, incubated at 100°C for 8 min, and then processed as above.

References

Aasen E, Medrano JF (1990) Amplification of the ZFY and ZFX genes for sex identification in humans, cattle, sheep, and goats. Biotechnology 8:1279–1281

Buoncristiani M, von Beroldingen C, Sensabaugh GF (1990) Effects of UV damage on DNA amplification by the polymerase chain reaction. In: Polesky HF, Mayr WR (eds) Advances in Forensic Haemogenetics, vol 3. Berlin: Springer-Verlag, pp. 151–153

Friedberg EC (1985) DNA Repair. New York: WH Freeman

Gill P, Jeffreys J, Werrett D (1985) Forensic application of DNA fingerprints. Nature 318:577–579

Higuchi R, von Beroldingen C, Sensabaugh GF, Erlich H (1988) DNA typing from single hairs. Nature 332:543–546

Igo-Kemenes T, Horz W, Zachau HG (1982) Chromatin. Annu Rev Biochem 51:89–121

Irwin D, Kocher TD, Wilson AC (1991) Evolution of the cytochrome b gene in mammals. J Mol Evol 32:128–144

Pääbo S, Gifford JA, Wilson AC (1988) Mitochondrial DNA sequences from a 7000 year old brain. Nucl Acids Res 16:9775–9787

Reynolds R, Sensabaugh GF, Blake ET (1991) Analysis of genetic markers in forensic DNA samples using the polymerase chain reaction. Anal Chem 63:2–15

Sambrook J, Fritsch EF, Maniatis T (1989) Molecular Cloning. A Laboratory Manual, 2nd ed. Cold Spring Harbor, N.Y.: CSHL Press

Sensabaugh GF, von Beroldingen C (1971) Genetic typing of biological evidence using the polymerase chain reaction. In: Farley MA, Harrington JJ (eds) Forensic DNA Technology. Chelsea, Mich.: Lewis Publishers, pp. 63–82

Walsh PS, Metzger DA, Higuchi R (1991) Chelex 100 as a medium for simple extraction of DNA for PCR-based typing from forensic material. Biotechniques 10:506–513

Ward WS, Coffey DS (1991) DNA packaging and organization in mammalian spermatozoa: comparison with somatic cells. Biol Reprod 44:569–574

Weye JS, Prestley LA, Budowle B, Shutler GG, Fourney RM (1989) A simple and sensitive method for quantifying human genomic DNA in forensic specimen extracts. Biotechniques 8:852–855

Witt M, Erickson RP (1989) A rapid method for detection of Y-chromosomal DNA from dried blood specimens by the polymerase chain reaction. Human Genetics 82:271–274

Dried Samples: Soft Tissues
10 DNA from Museum Specimens

ALAN COOPER

1. Introduction

The use of museum collections as a source of DNA offers many unique advantages. A diverse collection of taxonomically identified specimens located in one place creates a range of opportunities for evolutionary and ecological research while avoiding costly field studies. Recorded specimen sexes and collection dates enable population, ecological, pathological, and genetic studies to be calibrated with time offering valuable temporal evolutionary insights (Thomas et al. 1990). Many museum studies can complement molecular work; morphological studies can suggest phylogenetic relationships for molecular testing, and archaeological studies and carbon-dated specimens provide an important temporal and spatial framework for ancient DNA (aDNA) studies. The polymerase chain reaction (PCR) is revolutionizing the role of the museum in science by drastically enhancing the amount of information that can be obtained from museum collections. Although the DNA recoverable from these specimens is generally less than 500 base pairs (bp) in length, the ability of PCR to selectively amplify targeted sequences, and to jump damaged points in the DNA (Pääbo et al. 1989) permits larger regions of DNA to be amplified and sequenced. Consequently, a range of genetic material evolving at rates fast enough to distinguish between individuals, and slow enough to examine large scale systematic relationships, has become the latest tool for biological investigation of museum collections.

2. Collection of Samples from Museum Specimens

2.1 Background

Molecular analysis generally involves the destruction of material. It is important to minimise the conflict between this fact and the need to preserve museum specimens for other studies. It is vital that molecular workers communicate closely with museum curators and that all molecular work on valuable specimens is carefully considered and a coordinated approach taken. An idea with merit is that a standard protocol be devised for approaching museums to obtain ancient material. This would include credentials, an explanation of the study and likely results, the effects of the sampling technique on the specimen and arrangements to report the research. Since only a small amount of DNA is needed for PCR, one possibility is that excess extract from a valuable specimen is returned to the museum for storage until the need for additional molecular studies arises. Museums should also be encouraged to invest in −70°C storage facilities to house frozen tissue collections, since specimens sent to museums contain potential molecular and morphological information. The National Frozen Tissue Collection of the National Museum of New Zealand was instrumental in providing samples for the study described below.

Samples should be collected using standard aDNA sampling techniques involving sterile gloves, disposable blades and plastic tubes, and minimizing the handling of the specimen. It is also advisable to use breathing masks and to avoid talking over the specimen. At least two samples should be taken and processed and amplified separately to ensure reproducibility and the chance of detecting cryptic contamination.

2.2 Materials Used as Samples

Successful amplification and sequencing of DNA from ancient specimens depends on sampling material that contains usable authentic DNA, and which also has minimal, or no, contamination with nonauthentic, or exogenous DNA. When collecting samples from a museum specimen a further aim is to minimize the impact on the aesthetic and scientific value of the specimen. Unfortunately, the areas of the specimen that are macroscopically well preserved are also the most suitable for sampling, as the autolytic and postmortem degradation affecting macroscopic structure similarly affects DNA. These areas are generally the extremities of a specimen, which are removed from the effects of enteric microbes following death, and are also preferentially dried, limiting the time for degradation of DNA (Pääbo 1985). If postmortem changes have been limited by prompt dehydrative or similar preservative treatment, the location of the area to be sampled is less important. Organs such as the liver, where enzymatic activity is particularly high, should probably be avoided if higher molecular weight DNA is desired (Lee et al. 1991).

During 1990 I was involved in a study of the molecular relationships of the extinct New Zealand moa in collaboration with the late Professor Allan C. Wilson and Dr Svante Pääbo at the University of California at Berkeley. With their expertise the project was successfully extended to develop a molecular phylogeny of all the extant ratites using sequences of the mitochondrial ribosomal small subunit RNA (12S) gene. During the period at Berkeley I was able to examine a veriety of ancient specimens from museums; this chapter presents some of the results, including a summary of the study reported in Cooper et al. (1992), and subsequent studies at Victoria University of Wellington, New Zealand.

The ratites are a group of large flightless birds found on the landmasses formed by the breakup of the Mesozoic super-continent, Gondwanaland. The ratite birds are the ostrich of Africa, the rheas of South America, the emu and cassowaries of Australasia, the kiwi of New Zealand, and formerly the elephant birds of Madagascar and eleven species of moa in New Zealand, all now extinct. The closest relatives of the ratites are thought to be the South American tinamou, which share the primitive structure of their skulls. A contentious debate about the origins of the ratites has raged for more than a century with the evolutionary origins and monophyly of the kiwi and extinct moa central issues (Cracraft 1974; Sibley and Ahlquist 1981; Houde 1986). Since preserved moa tissue and bone samples were available, our aim at Berkeley was to regain preserved genetic information and address some of these issues.

During the ratite study, many materials were found to be suitable sources of DNA for PCR based analysis, including soft tissues, feather, eggshell, and bone. Approximately 0.1 g of dried soft tissue or 1–2 feathers generally provided ample DNA for PCR. The DNA containing material is found at the tip of the feather quill and if the specimen skin has shrunk during the preservation process, less dermal material will be present in a plucked quill and more feathers may be required. The amount of DNA obtainable from eggshell samples depends upon the amount of membrane remnants or attachment points (mamillary cones) preserved on the interior surface of the shell, as few cells appear to be inserted during the formation of the eggshell itself (Taylor 1970). The use of bone as a source of DNA is covered below and elsewhere in this volume.

A variety of techniques have been used to preserve museum specimens, many of which can adversely affect DNA contained within the specimen, or offer an additional source of contamination. Analysis of soft tissue remains of moa and other vertebrates indicate that naturally mummified specimens yield higher molecular weight DNA if the mummification process was prompt and if the specimen was not subsequently exposed to water. Specimens subjected to cleaning and preservative treatments yield varying amounts of DNA depending on the treatment. If the preservative technique involves a surface treatment, such as salting (e.g. with alum, borax, or saltpeter), coating (e.g. with varnish), or application of toxins (e.g. arsenic), the effects of the treatment can be minimized by obtaining material from a subsurface area. Hall et al. (1993) report that the inhibitory effects of alum can be removed by repeated washing with water. Coatings, such as the shellac encountered on one of the moa specimens (*Emeus*) can often be removed with a solvent (ethanol, Horie 1987). Solubilizing the coating can be inadvisable since this may cause increased penetration of the sample by the coating. The effects of preservative techniques involving immersion of the specimen in fixative are more difficult to avoid although bone may offer some protection. Museum records should be checked for details of the state of the specimen upon acquisition and subsequent preservative treatments.

Because museum specimens are handled during acquisition, preservation procedures, and display, their surfaces are more likely to be contaminated with human cell debris than other ancient samples. Consequently, it is desirable to take subsurface samples whenever museum related contamination and potentially damaging effects of surface chemical preservative treatments must be minimized. Bone can act as an extra barrier against museum contamination and possibly also against autolytic, oxidative, and hydrolytic damage (see below) and may be particularly suitable as a source of DNA when certain liquid based preservation techniques have been applied. Bone is also more commonly preserved than soft tissues, so specimens are often of less value to museum collections and the range of samples and DNA yield can be greater.

3. Extraction and Purification Procedures

To minimize contamination use a breathing mask and work in an area with low or filtered air movement. The outer surface of the sample should be cleaned of debris (ethanol can be used if the surface is relatively nonabsorbent) or abraded, if possible, with a scalpel or sandpaper, to remove residues of previous human cell contamination or preservative treatments.

The following extraction method (Thomas et al. 1990) was used in the ratite study. Tissue and feather samples were finely chopped on a clean surface (e.g. tinfoil) and incubated overnight with gentle agitation at 37–55°C in approximately 10 volumes of extraction buffer containing 10 mM Tris-HCl (pH 8.0), 2 mM EDTA, 10 mM NaCl, 0.5 mg/ml proteinase K, 10 mg/ml DTT, and 1% SDS. Temperatures approaching 55°C inhibit DNases but not the action of proteinase K. An extraction without any sample should always be performed as a control to test for contamination of the reagents.

DNA was successfully extracted from swamp and cave preserved bones using a simple technique. Approximately 1 cm^2 of cortical bone is cut out with a handdrill and hacksaw blade. The outer surface is removed with sandpaper and the sample is wrapped in tinfoil and broken into small fragments with a hammer. The fragments can then be powdered under liquid air using an acid washed mortar and pestle or electric coffee grinder (although the latter has associated sterilization problems). The powder is partly decalcified in approximately 15 volumes of 0.5 M EDTA (pH 8.0) for 1–2 days and collected by centrifugation. The sediment is then incubated in extraction buffer as above. An alternative to the centrifugation step is to place the sediment and solution in preboiled Spectra/por-3 dialysis tubing (Spectrum Medical Industries, Los Angeles) following the decalcification and dialyze against 100 volumes of 10 mM Tris-HCl (pH 8.0) for 24 hr with three changes of buffer. The sample is then concentrated by placing the dialysis tubing in 20% polyethylene glycol, 10 mM Tris-HCl (pH 8.0) for several hours, before the dialysate is placed in extraction buffer as above. Both methods have been used successfully on subfossil bones.

The DNA is purified following the proteinase K treatment by two extractions with phenol and one with chloroform/isoamyl alcohol. The aqueous phase is concentrated and purified with sequential additions to Centricon-30 microconcentrators (Amicon Inc., Danvers, Mass.) to minimize loss of material. Isopropanol precipitation is less efficient and involves several steps and solutions which increases the risk of contamination.

The samples used in the ratite study are reported in Cooper et al. (1992) except *Emeus crassus* (bone NZM 469, National Museum of New Zealand). Two moa soft tissue specimens (Earnscleugh Cave *Emeus crassus* and Knobby Range *Dinornis novaezealandiae*, both of Otago Museum, New Zealand) had been coated with varnish. At least two independent extractions were performed for samples of ancient specimens.

4. PCR Amplification and Inhibitors

4.1 Amplification Techniques

The PCR reaction buffer given here is that of Pääbo (1990), with elevated levels of bovine serum albumin (BSA); 67 mM Tris-HCl (pH 8.8), 2 mM $MgCl_2$, 250 µM each of dGTP, dATP, dTTP and dCTP, 2 mg/ml BSA (Sigma, fraction V), and 1.25 units of *Taq* polymerase. A few microliters of DNA extract should be initially added to make a 25–50 µl reaction mixture. Following amplification and visualization with ethidium bromide, the amount of added template should be increased if primer dimers but no specific products are observed, or decreased if neither are observed (Pääbo 1990). The PCR thermal profile used for the 12S gene amplifications was denaturation at 92°C for 40 sec, annealing at 55°C for 1 min, and extension at 72°C for 1 min. It was found that 40 cycles of PCR were required for ancient extracts. All PCR reactions should be accompanied by extract and PCR control reactions to detect any laboratory contamination.

Initial PCR amplifications of 12S fragments from moa soft tissue remains were generally limited to less than 150 bp, as has been observed previously (Pääbo 1989). Due to the small size of the amplifications 10 primers were designed to amplify a series of overlapping regions creating a contiguous 400 bp sequence of the 12S gene. Subsequent work showed that soft tissues of well preserved moa specimens and extracts from bones often permitted amplifications of around 250 bp and at least 400 bp, respectively. The primers used were:

12SA (L1091)	5'-CAAACTGGGATTAGATACCCCACTAT-3'
12SB (H1478)	5'-GAGGGTGACGGGCGGTGTGT-3'
12SBm (H2148)	5'-GAGGGTGACGGGCGGTGTGTGCAT-3'
12SD (L1373)	5'-AGAAATGGGCTACATTTTCT-3'
12SE (L1873)	5'-ACCCACCTAGAGGAGCCTGTTC-3'
12SF (H2023)	5'-AGAAAATGTAGCCCATTTCT-3'
12SG (H1892)	5'-GGCAAGAGATGGTCGGGTGTA-3'

12SH (H1985)	5'-CCTTGACCTGTCTTGTTAGC-3'
12SI (L1944)	5'-TACATACCGCCGTCGCCAGCCC-3'
12SJ (L1999)	5'-CCCCCGCTAACAAGACAGGT-3'

The letters L and H refer to the light and heavy strands, and the numbers refer to the position of the 3' base of the primer in the complete chicken mitochondrial DNA sequence (Desjardins and Morais 1990). 12SA, 12SB, and 12SD were based on previously reported primers (Kocher et al. 1989; Thomas et al. 1989) and have numbers referring to the complete human mitochondrial sequence (Anderson et al. 1981).

When designing PCR primers, it is important to note that the 3' base is important for primer binding specificity (Kwok et al. 1990). Pääbo (1989) noted that for analysis of aDNA, the 3' base should not be designed to bind to a template thymine or cytosine residue, since these are preferentially degraded in aDNA. Kwok et al. (1990) also showed that a thymine residue should not be used in the primer 3' position if sequence specificity is required for that residue of the template (e.g. to prevent contaminating human sequences from being amplified). Conversely, if the template sequence is unknown, or if a universal primer is desired then thymine can be used in the 3' position.

The 3' bases of primer pairs should not be complementary for PCR amplifications of ancient or minute amounts of DNA, as this increases the chances of primer dimer formation and subsequent competition with the specific amplification product. This problem is exaggerated for PCR profiles with low annealing temperatures. If complementarity of primer 3' bases cannot be avoided, or primer dimer formation is a problem due to inefficient primer annealing or damaged template, biphasic PCR (Ruano et al. 1989) can increase the amount of specific product. In this method only one primer is present in the reaction mix for the first 5–10 cycles, permitting the construction of complete copies of one template strand via jumping PCR (Pääbo et al. 1989) or direct extension, but preventing primer dimer formation. The second primer is then added and the completed strand permits conventional PCR amplification. This technique was used successfully in amplifications of moa mitochondrial ND6 gene fragments to outcompete primer dimers formed because of complementary 3' bases.

The ratite 12S gene PCR products were isolated from a 4% NuSieve agarose gel and sequenced using unbalanced priming and direct sequencing. Double stranded and solid phase sequencing using biotinylated primers have been used with bone and ancient samples. Two independent extracts were performed for ancient samples and both strands of each were sequenced.

4.2 PCR Inhibitors

The BSA concentration was found to be critical for preventing inhibition of the PCR reaction for several modern samples and nearly all ancient samples. Ancient specimens preserved in a totally dry environment were the only exceptions. Heme and other porphyrin products, such as cytochromes, are known to inhibit PCR (Higuchi 1989) and aDNA extracts contain additional inhibitors (Pääbo et al. 1988; Hagelberg and Clegg 1991). While determining the concentration of BSA necessary to overcome PCR inhibitors present in extracts from moa bones (Fig. 1), the inhibition was seen to be related to the presence of water in the preservation environment of the bones. While hydrolytic damage of DNA is a component of the inhibition, there also appear to be water-soluble soil components that act as inhibitors which diffuse into the deposited bones. The severe inhibition of PCR amplifications of some DNA samples from moa bones in swamps suggests that plant cytochrome products in the swamp matrix may be responsible. In this way the environ-

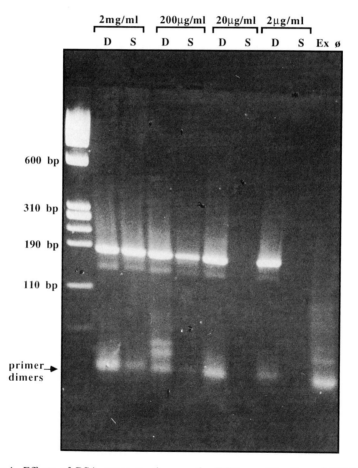

FIGURE 1. Effects of BSA concentration on the PCR amplification of 180-bp 12S DNA fragments from 3,000-year-old moa bones. The amplification of specific product and primer dimers shows that inhibitors present in extracts from bone preserved in an alkaline swamp (S) are overcome at high concentrations of BSA (2 mg/ml). The extracts from bone of similar age preserved in a dry cave (D) exhibit no inhibition.

ment of preservation may provide the specimen with an unique cocktail of PCR inhibitors, and consequently may be as important as the age of the specimen for the analysis of preserved DNA.

The CTAB plant DNA extraction method has been reported to reduce PCR inhibition of arthropod DNA extracts and this method also reduces the amount of co-purifying polysaccharides (Wilson 1988). Inhibitors of PCR were eliminated from bloodstain extracts through the use of a Chelex-100 based extraction procedure (Walsh et al. 1991). Chelex purification of ancient feather extracts has also been carried out with success (Ellegren 1991).

DNA extracts from frozen blood samples of the reptilian tuatara (*Sphenodon punctatus*) often proved difficult to amplify by PCR, and the inhibition was not reduced through the use of BSA. However, DNA extracts obtained from tuatara soft tissues (muscle) were easily amplifiable, demonstrating the possiblility of tissue specific inhibition.

5. Effects of Preservation Techniques on DNA

5.1 DNA Yield

For most aDNA studies the damaged template and exponential nature of the PCR reaction make it difficult to estimate the amount of template initially present. If the DNA is separated by agarose gel electrophoresis, crosslinks present in degraded DNA (Pääbo 1989) can cause the DNA to migrate at a speed consistent with higher molecular weight DNA. Hybridization studies of bone extracts have demonstrated that the majority of DNA visualized in this manner can be exogenous in origin, resulting from microbial contamination. Consequently the DNA molecular weight estimates in this study are based on the largest size fragment that could be amplified by PCR.

DNA extracted from moa bones was of considerably higher molecular weight than that extracted from surrounding soft tissues (Fig. 2). The relative DNA yield from specimens in this study in descending order, was from; dry preserved bone, swamp or partially moist preserved bone, naturally mummified tissue and artificially dried tissue, and chemically preserved tissue and naturally mummified tissue exposed to moisture during preservation. This suggests that bone is acting to shield DNA from environmental and organismic factors causing degradation and should be the preferred sampling material for ancient specimens, as suggested above. Considerably greater amounts of DNA have been obtained from spongy rather than compact bone (Lee et al. 1991) although this result was obtained for non-ancient specimens. The protective insulating properties of the denser cortical bone may be of greater importance for ancient samples.

Moa muscle, tendon, and skin samples that appeared to have been mummified slowly or exposed to moisture after deposition yielded DNA of generally similar molecular weight, ranging from 100 to 150 bp. The NZM S23808 *Megalapteryx* specimen which was mummified rapidly, provided material

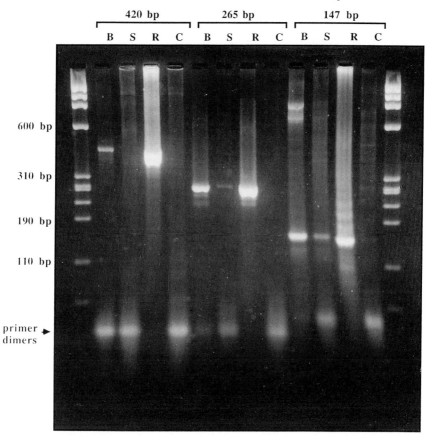

FIGURE 2. PCR amplification of 420-bp, 265-bp, and 147-bp 12S DNA fragments from 3,350-year-old moa bone (B), surrounding soft tissue (S), and modern rhea tissue (R) extracts and a PCR control (C). The bone extract permits higher molecular weight DNA amplifications than surrounding soft tissues. Primer dimers observed in the soft tissue amplifications indicate that the lack of specific product is not caused by PCR inhibition.

which had been exposed to varying amounts of moisture during preservation. Tissue in contact with the ground behaved similarly to the samples above, but tissue that had not been subsequently exposed to moisture permitted amplifications of up to 250 bp. Thomas et al. (1990) show similar results for specimens that had been dried quickly. Modern avian roadkills that have been naturally mummified, and museum feather samples up to 70 years in age, contain 12S DNA fragments of up to 400 bp. Preserved reptilian and avian specimens can be expected to contain higher amounts of DNA compared to mammalian samples due to the presence of nucleated erythrocytes in samples. 12S DNA amplifications were obtained from eggshells when dried membrane remnants were visible on the shell interior, but not when the shell was clean.

DNA extracted from shell will probably be maternal and not embryonic in origin (Taylor 1970).

Certain acid based tanning agents (e.g. vegetable tannins) appear to degrade DNA. Soft tissues from an extinct giant gecko specimen, *Hoplodactylus delcourti*, thought to have been tanned with vegetable-based tannins in the nineteenth century, did not yield DNA suitable for PCR amplification. X-ray analysis showed that limb bones had been left in situ during the preparative technique but unfortunately this material, which might provide better protection for DNA from the tanning, could not be obtained to conduct analysis.

Specimens treated with fungicidal or insecticidal agents such as thymol and napthol, yielded DNA which permitted amplifications of similar size to comparable untreated specimens.

5.2 *Contamination*

The *Emeus* and *Dinornis* specimens that had been coated with varnish possessed degraded DNA that would not routinely permit PCR amplifications over 80 bp, and gave inconsistent sequences with considerable human contamination. The contamination may have resulted from museum related handling or some process connected with the varnish. The shellac base used for the varnish coat on the *Dinornis* specimen is acidic (Horie 1987) and may have degraded the *Dinornis* DNA. Alternatively, human cell debris deposited with the varnish may have been the source of contamination, outcompeting a low level of endogenous DNA. Shellac was introduced into museum conservation practices in 1868; it is identifiable as it oxidizes to an opaque black color. The coating on the Emeus specimen is unidentified but resembles dammar, a triterpenoid resin used since 1842 that is characterized by a hard, brittle, yellow or golden coating (Horie 1987). The use of varnish is often associated with excessive degradation of the specimen prior to museum collection and consequent low levels of endogenous DNA.

During unsuccessful experiments to amplify DNA from bones of the Madagascan elephant bird, a low level of modern rhea DNA contaminant was detected in the PCR nucleotide mix. Although one of the components was known to be contaminated, the extract and PCR controls appeared negative. Amplifications of several elephant bird bone extracts were always positive, as well as contaminated, suggesting that the amplification of low copy number contaminant in the reactions was not purely random (Fig. 3). It appears as though the contamination of aDNA amplified by PCR can exhibit some form of carrier effect whereby a component present in some extracts can prevent inhibition, or facilitate the amplification, of single or a few copies of contaminating DNA. Control extracts that lack this agent do not amplify the low level contaminant, giving false negatives. This potential hazard should be explored more fully, and may be of particular concern in studies of ancient human DNA where the contamination is cryptic or otherwise difficult to identify.

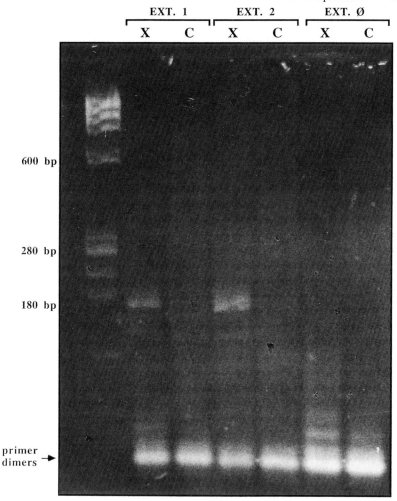

FIGURE 3. PCR amplifications of 180-bp 12S DNA fragments from two Madagascan elephant bird extracts with contaminated (X) and clean (C) PCR components. Reactions with clean PCR components give no amplification, indicating there is no endogenous DNA source. The nucleotide stocks in all contaminated reactions have trace amounts of modern rhea DNA, yet only the reactions with extract template give positive (contaminated) results. The extract control, which contains no organic material, does not permit amplification of the contaminating modern DNA, giving a false negative signal.

Cryptic contamination is a significant problem for museum collections, since related specimens are often stored or treated together and the contaminating agent is likely to be DNA from the same or a closely related species. Several amplifications of 12S DNA from the *Emeus* specimen were contaminated with chicken DNA, and the chicken sequence was similar enough that it was not immediately identified. Eventually it was noticed that a large portion of the considerable sequence differences between *Emeus* and other moa specimens was concentrated over a small segment of sequence. This region was located between the 12SE and 12SF primers which had been used to amplify a (contaminated) internal segment of the 12S sequence. Unequal distribution of sequence changes should be regarded with suspicion, although it can be caused by constraints imposed on the molecule (Irwin et al. 1991).

Where DNA sequences are ambiguous due to sequencing signals occuring for more than one nucleotide (multiple sequences), the sequences can often be used to determine the contaminating species. For example, human 12S sequences observed in varnished moa specimens may have been introduced during the coating treatment. Museum specimens treated with glue (e.g. eggshells, bones) could similarly be expected to yield human sequences, or horse, cow or pig sequences if the glue was of animal origin. Other preservative materials that are possible sources of exogenous DNA include varnishes and gelatine, casein, and bee-wax derivatives.

6. Sequence Analysis

Sequences of approximately 390 bp of the 12S RNA gene from extant and extinct ratites (Fig. 4) were aligned using the computer program ESEE (Cabot and Beckenbach 1989) and compared with the prokaryotic small ribosomal subunit secondary structure model (Neefs et al. 1990). The base compositions of the sequences were statistically indistinguishable. As Figure 4 shows, sequences adjacent to insertion/deletion sites were excluded from analysis because ambiguous homologies made alignment difficult (Cooper et al. 1992). The ratite data were subjected to parsimony, maximum likelihood, neighbour-joining, and Hadamard analyses (see Penny et al. 1992). Parsimony analysis was performed using the bootstrapping function of PAUP 3.0g (Swofford 1989) and the topology of parsimony trees constructed from 1,000 bootstrap replications analyzing the 92 variable positions (containing 69 informative sites) is shown in Figure 5. Bootstrap values over 95% are highlighted in the figure. Maximum likelihood analysis (Cooper et al. 1992) and neighbour-joining analysis (data not shown) give the same topology. Hadamard analysis (data not shown) suggests the tree is robust, and that there is no signal linking the kiwi and moa as a monophyletic clade.

The evolutionary trees produced by phylogenetic analysis of the 12S sequences show with high confidence that the kiwi and moa are not a monophyletic group and that the kiwi is genetically closer to an Australasian clade of ratites containing the ostrich (formerly widely distributed over Asia, Olson 1985), emu, and cassowary. The moa and rhea clades are both monophyletic and appear to have diverged early within the ratites. This is consistent with paleontological evidence of a rhea ancestor present in the late Paleocene of Brazil (Alvarenga, 1983). When the tinamou data are added they connect to the tree at the indicated root.

L.SP. KIWI GCTTAGCCCTAAATCCTGGTGCTTACATTACCTTAGTACCCGCCGCCCGAGAACTACGAGCACAAACGCTTAAAACTCTAAGGACTTGGCGGTGCCCTAAACCCACCTAGAGGAGCCTGTTCTATAATCGATAACCCACGATA
ROA KIWI   ..................A...G..................
BROWN KIWI ....................G.....C..............
CASSOWARY   .......T.A.A.CG.TA.....G..T..............
EMU   .......T.A.A.C.C....AG..C.T....T.........
OSTRICH   ..T.....................................
ANOMALOPTE .......A.A.A....CCC..AC..T..........C....
PACHYORNIS ...........A.A.A...CC..AC..T........C....
EMEUS   ...........A.A.A...CC..AC..T........C....
DINORNIS   .G................TA.A.A...CC..AC..T.....C
MEGALAPTER ...........A.A.A...CCC.AC..T........C....
LESSER RHE ...........C.A.A...CCC..C..T........C....
GREATER RH .......TA.A.A...A.C..C..T..........CT....
TINAMOU   .......T...T....T....

L.SP. KIWI CACCCAACCATCCTTGCCAA-CACAGCCTATATACCGCCGTCGCCAGCTCGCCTAAAATGAGAGACAACACAGCGAACAACACAGCTATC-CCAGCTAACAAAGACAGGTCAAGGTATAGCCTATGGAGATGGAAGAAATGG
ROA KIWI   ..................................C...........................CA......C.C..G
BROWN KIWI ...........TT..............C...........A.......G..G.T...C.C--T.C.........C.C.
CASSOWARY   .G.........................C...........A.......T.......G..T..C.C--TTC.......G.
EMU   .......C.C...T...........C...........A.......T..G..T..CCA--.C....A.....G.
OSTRICH   .G........G....-T.........C................................T..CC-...C
ANOMALOPTE .......G......C.C.CG-TG....C...........A.......T..G.......CTCA..C........A.......GAG
PACHYORNIS .......G.C.C...C-TG....C...........A.......T..G..........C.......A.
EMEUS   .......G.C.C...C-TG....C...........A.......T..GG.......C.C--..A.
DINORNIS   .......G.......C-TG....C...........A.......T..G.......C.C--TC.A.
MEGALAPTER .......G.......C-TG....C...........T.G......T..G.......C.C--...A.
LESSER RHE .......G.......C-TG....C...........T......A.G.T.T.......C........A.
GREATER RH .......G...C-TG....C...........T.G..A..AGG.T.T.......C..........A..G
TINAMOU   T.......G.......CC-T.......C.......A.CCAC......A..G.T..CC---..T.........AC...C.C

L.SP. KIWI ----ACGAAAAGGGTGTGAAATCCCCCTTAGAAAGGCGGATTAGCAGTAAAACAGAATAAGAGAGTCTTATTTAAG-CTGGCCCTAGGGC
ROA KIWI   -----A..............................A.........................A...
BROWN KIWI -----CC...................A.............................A--.C..
CASSOWARY   -----CC..A--.C..
EMU   -----TCC........AT......................A-.C..T.
OSTRICH   -----AC.........G.A.A....A.T.C..............A-.C..T..TC..T
ANOMALOPTE -----C..CC.......G..A.A....A.T..C.A...............C...AT.C
PACHYORNIS -----C..CC.......G..AAGA....CT.T..C..............T...C..AT.C
EMEUS   -----C..CC.......G..AAG....C..T..C..............T...C..AT.C
DINORNIS   -----C..CT......G..AAG....C.T..C............T.G.C..AC.C.
MEGALAPTER -----C..CC.......G..AAGA....C..T..C..............T.G.C..AC.C.C....G
LESSER RHE -----GTC..........GTG..G..C..A...C.C.............GTG.G..C.A.C.C.C
GREATER RH -----TC..C.ACC----........CA...C-A....CC...........GT.G..C.A.C.C.C...-.C
TINAMOU   -----CC..C.ACCTTTACTTATCTTT..G..G..TAA.....CG...........CTG...CT.A.............A--.T

L.SP. KIWI GCTACAATTTCTAAAATAGAAATACT

FIGURE 5. Parsimony phylogenetic tree produced by bootstrap analysis of ratite 12S sequences. Bootstrap values are shown above the branches, and branch lengths are proportional to the average number of transversions between taxa linked by a given node.

The positions of the kiwi and ostrich within the Australasian clade stimulate new hypotheses about ratite evolution. The ostrich is known to have been formerly widespread in Asia and has been suggested to have evolved there and migrated to Europe before recently being restricted to Africa (Olson 1985; Houde and Haubold 1987). Audley-Charles (1987) reviewed the migration of flora between Gondwanaland and Asia, and suggested that island chain routes existed between the remnant Gondwanaland landmass and Asia over a wide period of time, including the Early Tertiary. Although it is premature to make firm conclusions, a hypothesis consistent with the paleontological and molecular data is that the ancestor of the ostrich diverged from the Australasian ratite clade on the remnant Gondwanaland landmass during the Early Tertiary. The ostrich ancestor migrated via island chains to Asia where it subsequently speciated and spread into Europe before eventually becoming restricted to Africa.

The relatively recent divergence of the kiwi within the ratite phylogeny is difficult to reconcile with the 80 Mya separation of New Zealand from Australia and Antarctica (Mayes et al. 1990). Island hopping and swimming have previously been suggested as possible dispersal methods for the ancestor of the kiwi (Sibley and Ahlquist 1981). Traditionally proposed island dispersal routes involving the Norfolk Ridge and New Caledonia (Fleming 1979) have recently been reassessed. Attention is currently centered on the role of the Tonga-Kermadec ridge (formerly aligned along the position of the current Norfolk Ridge) in the dispersal of New Zealand taxa. It is also possible that the ancestor of the kiwi was flighted (Houde 1986; Cooper et al. 1992), and if this is correct, it appears that the various ratite lineages have lost the power of flight independently.

Further analysis of ratite evolution using mitochondrial and nuclear DNA sequences is currently under way. These studies are designed to increase the resolution of the separation dates of the Australasian ratite birds, and generate a complete phylogeny of the moa species. The ratite study illustrates the possibilities museum specimens offer as sources of DNA. Although certain preservative techniques appear detrimental to the preservation of DNA, and preventing contamination can be difficult, if these problems can be overcome then museum specimens offer unique opportunities to the evolutionary biologist.

Acknowledgments. I would like to thank the many museum staff who have generously given their time, advice, and precious specimens for analysis, particulary P. Millener, T. Worthy, J. Darby, and staff of the National Museum of New Zealand. I gratefully acknowledge the assistance of A.N. Baker, G.K. Chambers, D. Penny, and the members of the Wilson lab, particularly Tom Quinn, whom the ratite project was stolen off. I acknowledge the helpful critical analysis of W.M. Lento, R.A. Cooper, and A.C. Stone and the financial support of the National Museum of New Zealand, Victoria University, the Royal Society of New Zealand, and the New Zealand Royal Forest and Bird Protection Society in this work. I remain permanently indebted to S. Pääbo and Allan C. Wilson. This chapter is dedicated to the memories of Allan C. Wilson and Jan Drga, two people without whom it could not have been written.

References

Alvarenga H (1983) Uma ave ratitae do Paleoceno Brasileiro. Bol Mus Nac Geol 41:1–7

Anderson S, Bankier AT, Barrell BG, de Bruijn MHL, Coulson AR, Drouin J, Eperon IC, Nierlich DP, Roe BA, Sanger F, Schreier PH, Smith AJH, Staden R, Young IG (1981) Sequence and organization of the human mitochondrial genome. Nature 290:457–465

Audley-Charles MG (1987) Dispersal of Gondwanaland: Relevance to evolution of the angiosperms. In: Whitmore TC (ed) *Biogeographical Evolution of the Malay Archipelago.* Oxford: Clarendon Press, pp. 5–25

Cabot EL, Beckenbach AT (1989) Simultaneous editing of multiple nucleic acid and protein sequences with ESEE. Comput Appl Biosci 5:233–234

Cooper A, Mourer-Chauviré C, Chambers GK, von Haeseler A, Wilson AC, Pääbo S (1992) Independent origins of New Zealand moas and kiwis. Proc Natl Acad Sci USA 89:8741–8744

Cracraft J (1974) Phylogeny and evolution of the ratite birds. Ibis 116:494–521

Desjardins P, Morais R (1990) Sequence and gene organization of the chicken mitochondrial genome: A novel gene order in higher vertebrates. J Mol Biol 212:599–634

Ellegren H (1991) DNA typing of museum birds. Nature 354:113

Fleming CA (1979) *The Geological History of New Zealand and its Life*. Auckland: Auckland University Press

Hagelberg E, Clegg JB (1991) Isolation and characterization of DNA from archaeological bone. Proc R Soc Lond B 244:45–50

Hall LM, Ashworth C, Bartsiokas A, Jones DS (1993) Experiments on inhibition problems in old tissues. Ancient DNA Newsletter 1(2):9–10 Roy Zoo Soc Lon

Higuchi R (1989) Simple and rapid preparation of samples for PCR. In: Erlich HE (ed) *PCR Technology: Principles and Applications for DNA Amplification*. New York: Stockton Press, pp. 31–38

Horie CV (1987) *Materials for Conservation: Organic Consolidants, Adhesives and Coatings*. London: Butterworths

Houde P (1986) Ostrich ancestors found in the Northern Hemisphere suggest new hypothesis of ratite origins. Nature 324:563–565

Houde P, Haubold H (1987) *Palaeotis weigelti* restudied: a small Middle Eocene ostrich (Aves: Struthioniformes). Palaeovertebrata 17:27–42

Irwin DM, Kocher TD, Wilson AC (1991) Evolution of the cytochrome *b* gene of mammals. J Mol Evol 32:128–144

Kocher TD, Thomas WK, Meyer A, Edwards SV, Pääbo S, Villablanca FX, Wilson AC (1989) Dynamics of mitochondrial DNA evolution in animals: Amplification and sequencing with conserved primers. Proc Natl Acad Sci USA 86:6196–6200

Kwok S, Kellogg DE, McKinney N, Spasic D, Goda L, Levenson C, Sninsky JJ (1990) Effects of primer-template mismatches on the polymerase chain reaction: Human immunodeficiency virus type 1 model studies. Nucl Acids Res 18:999–1005

Lee HC, Pagliaro EM, Berka KM, Folk NL, Anderson DT, Ruano G, Keith TP, Phipps P, Herrin GL, Garner DD, Gaensslen RE (1991) Genetic markers in human bone. I. Deoxyribonucleic Acid (DNA) Analysis. J For Sci 36:320–330

Mayes CL, Lawver LA, Sandwell DT (1990) Tectonic history and new isochron chart of the South Pacific. J Geophys Res 95:8543–8567

Neefs JM, Van de Peer Y, Hendriks L, De Wachter R (1990) Compilation of small ribosomal subunit RNA sequences. Nucl Acids Res 18:2237–2317

Olson SL (1985) The fossil record of birds. In: Farner DS, King JR, Parkes KC (eds) *Avian Biology: Volume VIII*. Orlando: Academic, pp. 79–238

Pääbo S (1985) Molecular cloning of ancient Egyptian mummy DNA. Nature 314:644–645

Pääbo S (1989) Ancient DNA: Extraction, characterization, molecular cloning, and enzymatic amplification. Proc Natl Acad Sci USA 86:1939–1943

Pääbo S (1990) Amplifying ancient DNA. In: Innis MA, et al. (eds) *PCR Protocols: A Guide to Methods and Applications*. San Diego: Academic Press, pp. 159–166

Pääbo S, Gifford JA, Wilson AC (1988) Mitochondrial DNA sequences from a 7,000-year old brain. Nucl Acids Res 16:9775–9787

Pääbo S, Higuchi RG, Wilson AC (1989) Ancient DNA and the Polymerase Chain Reaction: The emerging field of molecular archaeology. J Biol Chem 264:9709–9712

Penny D, Hendy MD, Steel MA (1992) Progress with methods for constructing evolutionary trees. TREE 7:1–12

Ruano G, Kidd KK (1989) Biphasic amplification of very dilute DNA samples via "booster" PCR. Nucl Acids Res 17:5407

Sibley CG, Ahlquist JE (1981) The phylogeny and relationships of the ratite birds as indicated by DNA–DNA hybridization. In: Scudder GGE, Reveal JL (eds) *Evolution Today. Proc. 2nd Intern. Congr. Syst. Evol. Biol.* Pittsburgh, P.A.: Carnegie-Mellon University, pp. 301–335

Swofford DL (1989) PAUP: *Phylogenetic Analysis Using Parsimony*, version 3.0 g. Champaign, Ill.: Illinois Natural History Survey

Taylor TG (1970) How an eggshell is made. Sci Am 222:89–95

Thomas RH, Schaffner W, Wilson AC, Pääbo S (1989) DNA phylogeny of the extinct marsupial wolf. Nature 340:465–467

Thomas WK, Pääbo S, Villablanca FX, Wilson AC (1990) Spatial and temporal continuity of kangaroo rat populations shown by sequencing mitochondrial DNA from museum specimens. J Mol Evol 31:101–112

Walsh PS, Metzger DA, Higuchi R (1991) Chelex® 100 as a medium for simple extraction of DNA for PCR-based typing from forensic material. Biotechniques 10:506–513

Wilson K (1988) Preparation of genomic DNA from bacteria. In: Ausubel FM et al. (eds) *Current Protocols in Molecular Biology.* New York: Wiley, sec. 2.4.1–2.4.5

Dried Samples: Soft Tissues
11 DNA from Herbarium Specimens

JOHN W. TAYLOR and ERIC C. SWANN

1. Introduction

Old, soft tissues from plants and fungi may be derived from many sources, including fossil beds (Golenberg et al. 1990; Soltis et al. 1992) and sites of human or animal habitation (Rogers and Bendich 1985), but by far the most voluminous sources of these tissues are dried herbarium collections. Owing to the importance of herbaria, this review of methods for the analysis of DNA in preserved plants and fungi, and of methods for their preservation for future DNA extraction, will focus on herbarium specimens. Much of the literature on DNA preservation in herbarium material has been concerned with the large amounts of DNA needed for direct visulization in electro-phoretic gels, or indirect visualization by hybridization to labeled probe DNA. Here, however, we will emphasize use of the polymerase chain reaction (PCR; Mullis and Faloona 1987) to amplify DNA extracted from small amounts of herbarium material, because this approach makes much more sparing use of herbarium material and because we have direct experience with PCR amplification of DNA from herbarium specimens (e.g., Bruns et al. 1990; Swann et al. 1991).

1.1 What Type of Material Is Found in Herbaria?

Herbaria typically contain preserved plants, fungi, and algae. Our own experi-ence in molecular evolution studies is with fungi, and there is a rich literature on methods for higher plant molecular systematics, so we will emphasize these two groups of organisms. The methods presented here, however, are not restricted to higher plants and fungi but should be equally applicable to mosses, liverworts, algae, and lower vascular plants. Herbarium material is usually dried and stored flat on herbarium sheets or in folded packets or small boxes. Prior to storage, the fresh or newly dried material may have been treated with solvents or pesticides. Less commonly, material is stored in liquid in sealed jars. The liquid may contain acid, aldehydes, alcohols or other organic solvents. We will review what is known about the effect of different storage methods on subsequent DNA extraction, a subject that has generated considerable interest (Pyle and Adams 1989; Doyle and Dickson 1987; Miller 1990).

1.2 What Sort of Question Can Be Addressed with Herbarium Material?

Any question of evolutionary biology that can be studied with fresh material can also be examined with herbarium material. Phylogenetic surveys can be broadened by examination of exotic taxa held in herbaria. Herbarium specimens can provide a sample of old populations for studies of changing genetic structure. Disputes about nomenclature might be settled by examination of DNA from type specimens. Researchers interested in the evolution and function of genes, rather than organisms, should also find herbaria useful for comparative investigations. The utility of DNA extracted from herbarium specimens is not limited to evolutionary questions: natural products researchers might screen herbarium specimens for variations on genes of interest just as they screen fresh collections for gene products.

The impact of nucleic acid evolutionary biology on herbaria is complex. On the one hand, the influence of prominent herbaria was reduced when researchers without immediate access to large collections began to carry out phylogenetic analyses based solely on DNA extracted from fresh specimens. But at the same time, the collections held in herbaria are more valuable now since we know that the genomes of specimens can be preserved along with the cell walls. This has certainly made it easier for curators to justify the role of herbaria to their physiologist and biochemist colleagues, as well as to administrators of government agencies supporting scientific endeavors.

1.3 What Questions Should Herbarium Curators Be Asking?

The challenge to herbaria is to provide molecular evolutionary biologists access to specimens without having them destroyed with crude extraction methods or contaminated with extraneous DNAs. This will be most difficult with old and rare specimens, where the curator must balance the need for immediate information with the knowledge that extraction methods are certain to improve in the future. The methods described below will do no more damage to herbarium specimens than do morphological examinations of rehydrated flowers or anatomical studies of small fungal fruiting bodies. Curators must take the lead in studying how best to preserve DNA in herbarium specimens, how to avoid contamination of old specimens with new organisms or new DNA, and how to curate surplus DNA extracted from existing specimens as well as DNA extracts included with new accessions.

2. Preservation Conditions

DNA in condition good enough for PCR amplification (Saiki et al. 1988) and subsequent nucleotide sequencing has been obtained from 20-million-year-old leaves of *Magnolia* (cf. Golenberg et al. 1990) and *Taxodium* (Soltis et al. 1992). Conversely, intact DNA may be difficult to find in leaves stored for half a week in acid-aldehyde fixatives (Pyle and Adams 1989). As Rogers and Bendich (1985) noted, "the extent of DNA degradation for the herbarium specimens appeared to be related to the condition of the leaf rather than the year in which it was dried."

2.1 Preservation Methods Encountered in Old Material

In general, old, air-dried material that has neither been treated with chemical preservatives nor with high heat has the best chance of yielding useful DNA. Doyle and Dickson (1987) found that leaves of *Solanum glutinosum* preserved by drying at 42°C overnight with subsequent storage at room temperature yielded much more high molecular weight DNA than leaves preserved by (i) fixation in formalin/acetic acid/ethanol (FAA; 70% ethanol:formalin:glacial acetic acid, 18:1:1), (ii) fixation in Carnoy's solution (ethanol:acetic acid, 3:1) followed by storage in 70% ethanol, or (iii) storage in 70% ethanol without fixation. Leaves dipped in 70% ethanol and then dried showed more degraded DNA than leaves dried without the ethanol dip. Of the chemical treatments, only chloroform:ethanol (4:3) preserved long DNA molecules, but just for a week. Freezing previously dried specimens at $-20°C$ did not affect the degradation of DNA. In conclusion, Doyle and Dickson point out that fresh material should be preferred. For example, they extracted enough high molecular weight DNA from a 12-year-old, air-dried *Solanum* leaf to see a typical restriction fragment pattern by hybridization to cloned ribosomal DNA (rDNA).

In a similar study, Pyle and Adams (1989) compared several different plant preservation methods. They found over a five-month period that the highest molecular weight DNA was extracted from spinach leaves that had been (i) frozen at $-20°C$, (ii) refrigerated at 4°C, or (iii) dried at 42°C, and stored either at room temperature or frozen. Leaves desiccated above anhydrous $CaSO_4$ or dried at room temperature and stored at room temperature had suffered DNA degradation after five months. DNA was seriously degraded in leaves preserved by (i) boiling in water, (ii) microwaving, or (iii) immersion in solutions containing alcohols, acids, aldehydes, glycerol, sodium azide, guanidine thiocyanite, pentachlorophenol, chloroform, NaCl, sodium hypochlorite, or trichloroacetic acid. Only a solution of 0.5 M EDTA provided any protection for the DNA, and then for less than one month.

With fungi, Haines and Cooper (1993) assayed the amount of DNA extracted from small cubes (4 mm on a side) of the mushroom *Agaricus brunnescens* that had been preserved by drying, refrigeration, or overnight immersion in various solutions. Again, air-dried specimens yielded the most DNA, with drying at 22°C yielding more than drying at 38°C. Refrigeration at 4°C for 12 hr or immersion in 95% ethanol for the same period were worse than drying. Immersion in 5% phenol or in 1% $HgCl_2$ was worse still, and 10% formaldehyde reduced the DNA yield to 15% of that of the air-dried material. Heating the mushroom cubes to 93°C made recovery of any DNA impossible.

In summary, with either plant or fungal herbarium material, air-dried specimens are most likely to contain useful amounts of high molecular weight DNA. Specimens stored in alcohols, aldehydes, or acids are likely to be disappointing. If, however, alcohol-preserved material is all that is available (cf. Grody, Chapter 5), PCR amplification may still be successful, especially with multicopy genes such as the nuclear rDNA repeat unit or organelle-encoded genes.

2.2 Preservation Methods to Be Used with Newly Deposited Material

The studies of herbarium preservation methods summarized above clearly show that dried material is preferable to material stored in liquid when it comes to DNA content. However, humid field conditions and a lack of drying facilities may make air drying impossible. In this regard it is noteworthy that two groups have studied heat-free drying methods using desiccants and found that DNA is well preserved.

Liston et al. (1990) have tested silica gel as a desiccant for spinach leaves and also tested anhydrous $CaSO_4$ as a desiccant for field-collected plants in Xinjiang, China. DNA extracted from air-dried spinach leaves was slightly better preserved than DNA taken from silica gel desiccated leaves, although DNA extracted from plants preserved by both methods began to show degradation after five months of storage. Field desiccation with $CaSO_4$ yielded useful quantities of high molecular weight DNA and even active enzymes from a variety of annuals, perennials, succulents, and small shrubs (trees were not tested).

Chase and Hills (1991) have refined the silica gel desiccation technique for use in field collecting. They show that although less DNA overall is extracted from silica gel desiccated leaves than from fresh leaves, a substantial amount of DNA with the same high molecular weight as the DNA taken from fresh material remains in the desiccated material. They also show that this DNA can be digested with restriction endonucleases.

For newly collected material, immediate air drying at 42°C is as good as any method. If access to dry, hot air is impossible, pressed specimens should be preserved as well as possible, but some material should be desiccated in silica gel in the field for later DNA extraction. DNA extraction from the desiccated material should be accomplished as soon as laboratory facilities are available, because DNA degradation will occur in the plant or fungal material over time.

3. DNA Extraction

The amount of template DNA needed for PCR amplification is so small that the simplest imaginable extraction methods will work. The rationale for all of these methods is to cause the cell walls and membranes of the organism to leak DNA into solution. Of the many possible methods, we give three: crushing, Chelex-100, and detergent.

3.1 Crushing

One simple method of obtaining DNA from a specimen is to crush a small quantity of material, thereby breaking the cell wall and allowing the DNA to go into solution (Swann et al. 1991). This method has been useful in DNA extraction for PCR amplification of fungal DNA from spores and fruiting bodies, and should be applicable to plant materials as well.

Although in one case a single spore has yielded sufficient DNA for PCR amplification of a multicopy gene (Lee and Taylor 1990), typical practice in the authors' laboratory is to use 5–10 large spores (e.g., Erysiphales ascospores) or ca. 100 small spores (e.g., rust teliospores or urediniospores), or a single, small ascomycete fruiting body (perithecium or cleistothecium). The spores or fruiting body are crushed between two standard glass microscope slides which have been coated with 5% dimethyldichlorosilane in chloroform. The coated microscope slides can be treated to destroy possible contaminating DNA. This treatment may involve baking at 200°C for 1 hr, or dipping in 12 N HCl for 15 min followed by rinsing in distilled water. Neither heating nor acid treatment with HCl will remove the silane coating, although strong base or chromic H_2SO_4 would do so.

After crushing, the material is retrieved by pipetting it up in a small volume (20 µm) of liquid. Low-EDTA TE buffer (10 mM Tris-HCl/0.1 mM EDTA, pH 8.0) with 1% β-mercaptoethanol has been used successfully. The buffer and crushed material forms a bead on the silane-coated glass, which facilitates DNA recovery. After incubation in closed tubes at 70–80°C for 15 min, the β-mercaptoethanol is allowed to evaporate in a vacuum oven at 50°C. Being careful to leave all cell debris behind, the liquid is removed from the tube and diluted. Aqueous dilutions of 10^3- or 10^2-fold are routinely used. With very small amounts of material the entire crushed spore can be added to the amplification reaction. The ability to amplify template DNA obtained by crushing tends to decline over time, even when the DVA is stored at −20°C over the course of a single month. DNA preparations from crushed material can yield sufficient template for many PCR reactions (as many as 400), but our successful amplifications have all been done with multicopy genes and within a short period of time following DNA extraction.

3.2 Chelating Agents

One consideration in preparing DNA is to inhibit the activity of DNA-degrading enzymes. Use of a chelating agent, such as the resin Chelex-100 (biotechnology grade, 150–300 μm bead size, Bio-Rad Laboratories, Richmond, Cal.), to deprive such enzymes of ions necessary for their function is a way to avoid this problem (Walsh et al. 1991). Chelex-100 has been used successfully for DNA extraction from fungi in our laboratory. A small quantity of material (the less material the better; a single rust telium or a few lichen apothecia are plenty) is placed in 5% aqueous Chelex and can be treated in various ways. Most procedures involve heating the sample to between 50°C and 100°C for from 15 min to 8 hr. Alternating rounds of freezing and thawing have also been used. An extreme approach is to autoclave the material in Chelex for five minutes at 120°C. The autoclaving procedure is successfully used with fresh material, but may be too extreme for herbarium specimens with DNA of dubious integrity. The DNA obtained can be stored frozen in the Chelex solution, but PCR should be performed quickly as degradation will occur over time. Again, although enough DNA is obtained with the Chelex preparation for many PCR amplifications, our successes have all been with multicopy genes and within a short time following DNA extraction.

3.3 Detergent Extraction

Extraction with detergents can be used to obtain DNA both for PCR and for techniques which require much larger quantities of DNA, such as restriction enzyme digestion. Two detergents commonly used are SDS (sodium dodecyl sulfate) and CTAB (hexadecyltrimethylammonium bromide). The hot CTAB method (Doyle and Doyle 1987) is popular among plant workers and has been applied to fungal specimens as well. This method starts with 0.5 to 1.5 g of fresh tissue and yields high molecular weight DNA. Another CTAB method uses milligram amounts of material and has extracted DNA from fresh, herbarium, and mummified plant tissues (Rogers and Bendich 1985). The SDS method (Lee et al. 1988) has been used to obtain PCR template DNA from fungal and algal tissue. Compared to crushing in buffer and ion chelating methods, detergent extractions may yield the best results in terms of the "longevity" of the stored DNA.

4. PCR Amplification

Although it is possible to extract enough DNA from herbarium specimens to visualize the extracted DNA in ethidium bromide stained agarose gels, or to visualize restriction fragments of the extracted DNA by hybridization to labeled probe DNA, we do not recommened this approach because it consumes too much herbarium material. With the advent of PCR, DNA from as little as one nucleus can provide template for amplification even of single-copy genes (Li et al. 1988).

4.1 Amplification from Minute Amounts of Tissue

Each type of DNA extraction method described above, i.e., detergent minipreps, chelex micropreps and single propagule preps, can provide sufficient DNA for PCR amplification. For detergent minipreps, Liston et al. (1992) used 30 mg of leaves from herbarium specimens (ca. 1 cm^2 of leaf) and we routinely use from 30 to 70 mg of ground, dried fungus for PCR amplification. For Chelex micropreps, 5 to 30 mg of ground, dried fungus have yielded useful template DNA for PCR amplification. With single propagules, one to ten ascospores or ca. 100 rust spores have provided sufficient DNA for subsequent PCR amplification (Lee and Taylor 1990; Swann et al. 1991). A protocol for a sample PCR amplification from crushed propagules is given in Figure 1.

4.2 Target Molecules and Primers

Multicopy sequences that evolve in concert have been the target molecules of choice because their higher copy number facilitates PCR and because the chance of comparing paralogous molecules is reduced (cf. Doyle 1992). Genes in the rDNA repeat and regions of organelle genomes are those in current use. Fungal phylogenetic studies (reviewed by Bruns et al. 1991) have concentrated on the rDNA repeat unit and mitochondrial ribosomal DNAs. Primers for these regions are available (White et al. 1990). Plant phylogenetic studies have also exploited nuclear rDNAs and the chloroplast-encoded gene for the large subunit of ribulose-1,5-biophosphate carboxylase/oxygenase (rbcL). Primers designed for plant 18S rRNA are available (Hamby et al. 1988); primers designed by Gerard Zurawski for rbcL are reproduced here in Figure 2.

4.3 Amplification Conditions

The reaction conditions recommended by the Taq polymerase manufacturer should be strictly followed because different manufacturers produce different enzymes and use different storage buffers. As everyone experimenting with PCR will discover, amplification is sensitive to any variation in conditions. Innis and Gelfand (1990) discuss the effects of enzyme concentration, dNTP concentration, Mg^{2+} concentration, buffer constitution, annealing temperature, and reaction timing on amplification efficiency and fidelity. The fidelity of Taq polymerase DNA synthesis is very good under the right conditions (Innis and Gelfand 1990), but random errors can be expected to occur in individual amplified molecules. Direct DNA sequencing of the population

1. Crush 50 to 200 rust spores between two silane-coated microscope slides (see text for silane coating procedure).
2. Dilute the crushate 1:1,000 in distilled, filtered water.
3. Prepare the PCR in 500-µl microcentrifuge tubes.

PCR Parameters for Microprep DNA

Stock reaction component	Stock solution concentration	Volume	Final concentration
Water		29.6 µl	
Buffer	10X	10 µl	1X
dNTPs	10X	10 µl	1X (200 µM each dNTP)
Taq polymerase	5 units/µl	0.4 µl	0.02 units/µl
Primer 1	10 µM	5 µl	0.5 µM (50 pmoles)
Primer 2	10 µM	5 µl	0.5 µM (50 pmoles)
Pre-template volume		60 µl	
Template	1×10^{-3} dilution of crushed spores	40 µl	
Total Volume		100 µl	

4. Cover the reactions with one drop of mineral oil (Sigma, St. Louis, Mo.).
5. Place the reaction tubes in a thermal cycling machine (Perkin-Elmer, Norwalk, Conn.).

Thermocycling Parameters

90°C 1 min (denaturation)
52°C 1 min (annealing)
72°C 1 min (elongation)
(4-sec extension of elongation step
40 cycles

6. Assay the reactions by electrophoresis of 5 µl of PCR product in 1% Sea Kem agarose /2% NuSieve agarose (FMC, Rockland, Maine) in 1X TAE buffer followed by staining in ethidium bromide (0.5 µg/ml).

FIGURE 1. Protocol for PCR amplification from crushed propagules. Many variants of this protocol are possible. Most protocols combine equal volumes of template and concentrated reaction mix, our 60:40 combination simply moves some of the water from the dilute template to the reaction mix. We use 10X PCR buffer supplied by the manufacturer of the *Taq* polymerase. The composition of a typical PCR buffer is given in White et al. 1990. Two liters of 10X TAE buffer are made by mixing, 242 g Tris base and 34 g Na Acetate·3H₂O in 1800 ml of water, adding 40 ml of 0.5 M EDTA, bringing the solution to pH 8.1 with about 75 ml of glacial acetic acid, and adjusting the final volume to 2 l with water.

Z-1	ATGTCACCACAAACAGAAACTAAAGCAAGT
Z-1R	ACTTGCTTTAGTCTCTGTTTGTGGTGACAT
RH-1	ATGTCACCACAAACAGAAACTAAAGC
Z-153R	AGAAGATTCCGCAGCCACTGCAGCCCCTGCTTC
Z-234	CGTTACAAAGGACGATGCTACCACATCGA
Z-346R	AAATACGTTACCCACAATGGAAGTAAATAT
Z-427	GCTTATTCAAAAACTTTCCAAGGCCCGCC
S-523	AAACCAAAATTGGGATTATCCGCAAAAAATTA
Z-674	TTTATAAATCACAAGCCGAAACTGGTGAAATC
Z-674R	GATTTCGCCTGTTTCGGCTTGTGCTTTATAAA
Z-895	GCAGTTATTGATAGACAGAAAAATCATGGT
Z-895R	ACCATGATTCTTCTGCCTATCAATAACTGC
S-967	GGAGATCATATCCACTCCGGTACAGTAGTA
Z-1020	ACTTTAGGTTTTGTTGATTTATTGCGCGATGATT
Z-1020R	ATCATCGCGCAATAAATCAACAAAACCTAAAGT
Z-1204	TTTGGTGGAGGAACTTTAGGACACCCTTGGGG
Z-1204R	CCCTAAGGGTGTCCTAAAGTTTCTCCACC
Z-1264	GTAGCTTTAGAAGCCTGTGTACAAGCTCGTAA
Z-1375R	AATTTGATCTCCTTCCATATTTCGCA

FIGURE 2. These oligonucleotides were designed by Gerard Zurawski (DNAX Research Institute, Palo Alto, California) for PCR amplification and dideoxy sequencing of the large subunit, chloroplast coded, ribulose-1, 5-bisphosphate carboxylase/oxygenase gene (*rbcL*). The pimers represent the regions most conserved between maize and spinach sequences. Z indicates a maize sequence, S indicates a spinach sequence, RH indicates a rheum sequence. The primer numbers indicate the approximate location of the primer in nucleotides beginning with the ATG translation start codon. An R following the primer number indicates that the oligonucleotide will prime DNA synthesis toward the 5' end of *rbcL*; primers without an R prime DNA synthesis toward the 3' end of the gene.

of amplified molecules will mask random amplification mistakes, but it will also mask natural variation in these multicopy molecules.

4.4 Copurifying PCR inhibitors

The small-scale DNA isolation protocols suggested here do not produce pure DNA. With all of the methods, inhibitors of PCR amplification may copurify with the DNA. The main remedy for potential inhibitors is dilution: even the crushed DNA preparations may be diluted 10^2- or 10^3-fold prior to amplification. However, our experience is that unsuccessful PCR amplification is most often due to poor template DNA and not *Taq* polymerase inhibitors. If dilution would reduce the DNA content too far, purification and concentration can be accomplished by washing in centrifugal filtration units designed for this purpose (e.g., Millipore Ultrafree-MC, 100,000 NMWL, Bedford, Mass).

5. Documenting Variation in PCR-Amplified DNA

5.1 Restriction Fragment Length Polymorphism (RFLP)

Liston et al. (1992) demonstrated that PCR could be used to obtain chloroplast DNA from herbarium specimens in a study of *Datisca* spp (Datiscaceae). Their work builds on that of Arnold et al. (1991), who studied the population genetics of *Iris* species by using PCR to amplify the region of chloroplast DNA containing the *rbcL* gene. To obtain DNA from herbarium specimens, Liston et al. (1992) used 30 mg of leaves from dried herbarium specimens and the hot CTAB extraction protocol mentioned above (Doyle and Doyle 1987; Rieseberg et al. 1992). The PCR product was then digested with a variety of restriction endonucleases and both site changes and length mutations were inferred directly from photographs of restriction fragments electrophoresed in ethidium bromide stained agarose gels.

5.2 Restriction Site Mapping

An elegant extension of the RFLP method has been developed by Vilgalys and Hester (1990), who used PCR to rapidly map restriction sites on a DNA region (cf. Epplen Chapter 2). Their method is more informative than comparing restriction fragment lengths because the restriction site locations are mapped, and it avoids the tedium of DNA filter hybridization. With this method, they made a set of PCR products of increasing length, each of which had at one end of the molecule a common primer and at the other end one of a series of increasingly distant primers. By comparing restriction digests of the PCR product of the entire region with digests of each of the increasingly long PCR products, the location of the restriction sites was accurately mapped.

5.3 Nucleotide Sequencing

If knowledge of the behavior of individual nucleotides would be useful, as it is in phylogenetic studies, then the PCR products may be sequenced. Bruns et al. (1990) successfully sequenced a part of the mitochondrial large subunit rDNA from PCR-amplified DNA extracted from 16 herbarium collections. The DNA was obtained from 5 to 30 mg of ground mushrooms or other basidiomycete fruiting bodies by the detergent miniprep method. Symmetric, or double-stranded, PCR amplification provided template for asymmetric, or single-stranded, amplification, which in turn provided template for sequencing by the chain termination or dideoxy method.

5.4 Electrophoretic Demonstration of Single-Nucleotide Variation

Where allelic variation at specific loci is of interest, as in population genetic studies, the analytical method of choice should discriminate between alleles on the basis of a single nucleotide difference anywhere in the DNA fragment, but do so with the laboratory simplicity of gel electrophoresis. Two methods aim to meet this goal: denaturing gradient gel electrophoresis (DGGE; Myers et al. 1987) and single-strand conformation polymorphism (SSCP; Hayashi 1991). With DGGE, double-stranded PCR products of equal length are electrophoresed in polyacrylamide gels containing a gradient of increasing urea concentration while the gel apparatus is immersed in a constant temperature water bath. When the double-stranded DNA reaches the urea concentration necessary for its denaturation, its mobility decreases tremendously. It has been shown that molecules several hundred base pairs in length that differ at a single nucleotide position can be discriminated by DGGE (Sheffield et al. 1990).

SSCP analysis is similar to DGGE in that single-nucleotide differences can be recognized by differences in electrophoretic mobility. With SSPC, single strands are electrophoresed in neutral polyacrylamide gels and are visualized by silver staining. SSCP is logistically simpler than DGGE because SSCP does not require gradient gels, peristaltic pumps and constant temperature water baths. DGGE, which uses double-stranded DNA, has the advantage of allowing the researcher to make duplexes between alleles from different individuals. These heteroduplexes can be used to test whether fragments of identical mobility have identical sequences because mismatched nucleotides will strongly influence denaturation.

6. Modern Contamination of Herbarium Material

The great sensitivity of PCR, its ability to produce tremendous amounts of product DNA from extremely small amounts of template DNA extracted from similarly small amounts of herbarium material, can also be a curse when contaminating DNA is amplified.

6.1 Contaminating Organisms

We have personal experience with several cases where sequence analysis demonstrated that PCR product amplified from small amounts of one fungus turned out to be the DNA of other, contaminating fungi. For example, samples of parasitic fungi forming macroscopic fruiting bodies on leaves may include cells of microscopic yeasts living on the leaf surface in nature, as well as decay fungi that have grown on the leaves during preservation and storage. DNA extraction methods may be more effective for the contaminating organisms, or the PCR amplification reaction may favor their DNAs. Obviously, molecular results that seem inexplicable in light of morphological or physiological knowledge should be checked carefully. However, contamination is neither insurmountable nor inevitable; for instance, amplification from Miocene *Magnolia* and *Taxodium* leaves has been successful in the face of 17 to 20 million years of potential contamination, although the storage conditions must have been remarkable (Pääbo and Wilson 1991).

6.2 Contamination by Carryover of PCR Product DNA

An even more pernicious source of contamination of specimens, and of DNA extracted from specimens, is the highly concentrated PCR product amplified from the DNA of other specimens. Kwok and Higuchi (1989) discuss the magnitude of the problem and the measures needed to control PCR contamination. Solutions containing PCR products can be counted on to form aerosols during pipetting and other laboratory manipulations. Therefore, PCR products must not be brought into areas where herbarium specimens are stored, or where DNA is extracted from herbarium specimens. Mechanical pipettors and other laboratory apparatus used to handle PCR product must not also be used for the extraction of DNA from other specimens. The pipette tips used should not permit the PCR product to come in contact with the pipettor (suitable are, e.g., plugged tips or positive displacement tips). If the method of analysis is limited to double-stranded DNA, and if substitution of dUTP for dTTP does not pose a problem, a contamination-controlling strategy using uracil-DNA glycosylase to degrade previously amplified DNA is worth considering (Longo et al. 1990).

7. Emerging Considerations

7.1 Curation of DNA samples

Just as herbarium curators are newly concerned with how best to preserve the DNA in herbarium specimens, they must also worry about curating DNA extracted from and amplified from herbarium specimens (Reynolds and Taylor 1991). The quest for DNA from specimens also raises questions about the curation of rare or small specimens, where any sort of destructive analysis cannot be sanctioned. If history is a guide, DNA extraction methods should continue to improve and damage to specimens should be lessened. Currently, sufficient DNA template for PCR amplification of multicopy genes can be obtained from less material than in required for anatomical or morphological examination, i.e., 1 cm^2 of dry leaf, 5 to 100 spores, a small ascomycete fruiting body, or a small slice of mushroom. Although enough DNA is obtained for hundreds of DNA amplifications, in practice only one or two molecules will be amplified and the excess DNA will probably degrade before other amplifications can be attempted. At present, only the smallest amount of tissue sample should be taken and, despite the possibility of degradation, any excess DNA extracted still ought to be kept because amplification methods may improve. Curators may decide that DNA extraction is too damaging for certain specimens, or they may prefer to control the damage to specimens by isolating DNA in their own herbarium lab and then loaning this DNA, along with the specimen or instead of the specimen. Unfortunately, the most secure herbarium procedures for avoiding contamination cannot prevent sloppy practice by researchers working with loaned specimens. Researchers, in turn, will also be concerned with control of the process, and are likely to prefer their own DNA isolation technique and immediate amplification to avoid template DNA degradation. Protocol concerning these issues should be part of each herbarium's loan policy.

7.1.1 Specimen DNA

So little template DNA is required for PCR amplification that the researcher is likely to extract excess herbarium specimen DNA. This excess DNA should be curated so that future researchers can avoid damaging the herbarium specimen further, as well as to save effort. Our experience is that mini- micro-, and crushed propagule DNA degrades during storage in liquid, even at $-20°C$. Also, herbarium directors are not likely to be enthusiastic about the cost of low temperature storage and all low temperature storage is prone to occasional failure. Fortunately, dried DNA samples store very well. A few microliters of DNA sample pipetted onto a plastic film and allowed to dry, or larger samples precipitated and dried in microcentrifuge tubes, should be stable over long periods. There is no reason that these samples should not be stored with the herbarium specimens from which they were extracted. The curator must protect the specimens from decay associated with microbes or insects but should ascertain that steps taken to eliminate insects or decay organisms do not destroy DNA samples. Dry storage with occasional freezing to control insects should not harm the DNA. Wet, chemical treatments are to be avoided (see Section 2 above).

7.1.2 Amplified DNA

DNA amplified from herbarium specimens poses curatorial hazard because of the danger of contamination with PCR product. Although the most likely sites of contamination will be the herbarium laboratory spaces devoted to DNA isolation and amplification, the thought of spilling PCR product on unanalyzed specimens is not appealing. Even though it would be convenient if the PCR product could be stored with the herbarium specimen, that is too dangerous because even the outside of the tube or package is likely to be contaminated with PCR product. Here is another area where herbarium curators can make a contribution, by studying methods to safely store amplified DNA with herbarium specimens.

7.2 Ownership Issues

Free access by qualified researchers to collections is essential to science. Herbaria, however, which spend a great deal of time and money preserving collections, might justifiably lay claim to some of the commercial gain got from information stored in their specimens. The individual researcher who extracted the DNA and the information is also a factor in this equation, as is the availability of the results of analysis to other scientists. None of these issues is unique to ancient DNA studies, and the experience of microbial culture collections (e.g., ATCC, CBS, CMI) housing industrial strains or the electronic databases containing DNA sequence information (GenBank, EMBL, DDB) may be valuable to curators wrestling with these issues.

Acknowledgments. We thank John Haines, New York State Museum, Aaron Liston, Loren Rieseberg and Michael Hanson, Oregon State University and Rancho Santa Ana Botanical Garden, and Gerard Zurawski, DNAX, Palo Alto, for sharing preprints of manuscripts or unpublished information. The comments on earlier drafts made by Mary Berbee, Gerard Zurawski, John Haines, and Loren Rieseberg were particularly helpful. Preparation of his chapter was supported by NIH AI28545.

References

Arnold ML, Buckner CM, Robinson JJ (1991) Pollen-mediated introgression and hybrid speciation in Louisiana irises. Proc Natl Acad Sci (USA) 88:1398–1402

Bruns TD, Fogel R, Taylor JW (1990) Amplification and sequencing of DNA from fungal herbarium specimens. Mycologia 82:175–184

Bruns TD, White TJ, Taylor JW (1991) Fungal molecular systematics. Annu Rev Ecol Syst 22:525–564

Chase MW, Hills HH (1991) Silica gel: an ideal material for field preservation of leaf samples for DNA studies. Taxon 40:215–220

Doyle JJ (1992) Gene trees and species trees: molecular systematics as one-character taxonomy. Syst Bot 17:144–163

Doyle JJ, Dickson EE (1987) Preservation of plant samples for DNA restriction endonuclease analysis. Taxon 36:715–722

Doyle JJ, Doyle JL (1987) A rapid DNA isolation procedure for small quantities of fresh leaf tissue. Phytochem Bull 19:11–15

Golenberg EM, Giannasi DE, Clegg MT, Smiley CJ, Durbin M, Henderson D, Zurawksi G (1990) Chloroplast DNA sequence from a Miocene *Magnolia* species. Nature 344:656–658

Haines JH, Cooper CR Jr (1993) DNA and mycological herbaria. In: Reynolds DR, Taylor JW (eds) The Fungal Holomorph: Mitotic, Meiotic and Pleomorphic Speciation in Fungal Systematics. Wallingford: CAB International, pp. 305–315

Hamby RK, Sims L, Issel L, Zimmer E (1988) Direct ribosomal RNA sequencing: optimization of extraction and sequencing methods for work with higher plants. Plant Mol Biol Rep 6:175–192

Hayashi K (1991) PCR-SSCP: a simple and sensitive method for detection of mutations in the genomic DNA. PCR Meth Appl 1:34–38

Innis MA, Gelfand DH (1990) Optimization of PCRs. In: Innis MA, Gelfand DH, Sninsky JJ, White TJ (eds) *PCR Protocols: A Guide to Methods and Applications.* San Diego: Academic Press, pp. 3–12

Kwok S, Higuchi R (1989) Avoiding false positives with PCR. Nature 339:237–8 (erratum published in Nature 339:490)

Lee SB, Taylor JW (1990) Isolation of DNA from fungi and spores. In: Innis MA, Gelfand DH, Sninsky JJ, White TJ (eds) *PCR Protocols: A Guide to Methods and Applications.* San Diego: Academic Press, pp. 282–287

Lee SB, Milgroom MG, Taylor JW (1988) A rapid, high yield mini-prep method for isolation of total genomic DNA from fungi. Fungal Genet Newsl 35:23–24

Li H, Gyllensten UB, Cui X, Saiki RK, Erlich HA, Arnheim N (1988) Amplification and analysis of DNA sequence in single human sperm and diploid cells. Nature 335:414–417

Liston A, Rieseberg LH, Adams RP, Do N, Zhu G-1 (1990) A method for collecting dried plant specimens for DNA and isozyme analyse, and the results of a field test in Xinjiang, China. Ann Missouri Bot Gard 77:859–863

Liston A, Rieseberg LH, Hanson MA (1992) Geographic partitioning of chloroplast DNA variation in the genus *Datisca* (Datiscaceae). Plant Syst Evol 181:121–132

Longo MC, Berninger MS, Hartley JL (1990) Use of uracil DNA glycosylase to control carry-over contamination in polymerase chain reactions. Gene 93:125–128

Miller OK (1990) Use of molecular techniques, the impact on herbarium specimens and the preservation of related voucher materials. Mycol Soc Amer News 41:29–30

Mullis KB, Faloona F (1987) Specific synthesis of DNA in vitro via a polymerase-catalyzed chain reaction. Meth Enzym 155:335–350

Myers RM, Maniatis T, Lerman LS (1987) Detection and localization of single base changes by denaturing gradient gel electrophoresis. Meth Enzym 155:501–527

Pääbo S, Wilson AC (1991) Miocene DNA sequences—a dream come true? Curr Biol 1:45–46

Pyle MM, Adams RP (1989) *In situ* preservation of DNA in plant specimens. Taxon 38:576–581

Reynolds DR, Taylor JW (1991) DNA specimens and the 'International code of botanical nomenclature'. Taxon 40:311–315

Rieseberg LH, Hanson M, Philbrick CT (1992) Androdioecy is derived from dioecy in Datiscaceae: evidence from restriction site mapping of PCR-amplified chloroplast DNA fragments. Syst Bot 17:324–336

Rogers SO, Bendich AJ (1985) Extraction of DNA from milligram amounts of fresh, herbarium, and mummified plant tissues. Plant Mol Biol 5:69–76

Saiki RK, Gelfand DH, Stoffel S, Scharf SJ, Higuchi R, Horn GT, Mullis KB, Erlich HA (1988) Primer-directed enzymatic amplification of DNA with a thermo-stable DNA polymerase. Science 239:487–491

Sheffield V, Cox D, Myers R (1990) Identifying DNA polymorphisms by denaturing gradient gel electrophoresis. In: Innis MA, Gelfand DH, Sninsky JJ, White TJ (eds) *PCR Protocols: A Guide to Methods and Applications.* San Diego: Academic Press, pp. 206–218

Soltis PS, Soltis DE, Smiley CJ (1992) An *rbcL* sequence from a Miocene *Taxodium* (bald cypress). Proc Natl Acad Sci USA 89:449–451

Swann EC, Saenz GS, Taylor JW (1991) Maximizing information content of morphological specimens: herbaria as sources of DNA for molecular systematics. Mycol Soc Amer Newsl 42:36

Vilgalys R, Hester M (1990) Rapid genetic identification and mapping of enzymatically amplified ribosomal DNA from several *Cryptococcus* species. J Bacteriol 172:4238–4246

Walsh PS, Metzger DA, Higuchi R (1991) Chelex 100 as a medium for simple extraction of DNA for PCR-based typing from forensic material. BioTechniques 10:506–513

White TJ, Bruns TD, Lee S, Taylor JW (1990) Amplification and direct sequencing of fungal ribosomal RNA genes for phylogenetics. In: Innis MA, Gelfand DH, Sninsky JJ, White TJ (eds) *PCR Protocols: A Guide to Methods and Applications.* San Diego: Academic Press, pp. 315–322

Dried Samples: Soft Tissues
12 High-Fidelity Amplification of Ribosomal Gene Sequences from South American Mummies

PETER K. ROGAN and JOSEPH J. SALVO

1. Introduction

The condition of nucleic acids in mummified tissues depends on several variables, including the presence or absence of specific embalming treatments, the conditions of storage, and the natural chemical reactions that have occurred over time (Pääbo et al. 1989). These factors substantially reduce the yield of nucleic acids from ancient tissues; however, ancient DNA (aDNA) can be analyzed by amplifying specific sequences in vitro, with modifications of the polymerase chain reaction (PCR) (Pääbo et al. 1988; Rogan and Salvo 1991). Even when nucleic acids can be retrieved, manipulation of aDNA with molecular biological techniques is not always successful. Inhibitors of the PCR process that either copurify with or are components of the nucleic acid template have been shown to interfere with the processivity of the DNA polymerase used to amplify the template (Rogan and Salvo 1990).

Previous studies of nucleic acids derived from Atacamanian mummified tissues have shown that damaged nucleic acids, rather than soluble contaminants, prevent amplification (Rogan and Salvo 1990). Some of the chemical modification might be a consequence of oxidation of both deoxyribose and bases, although other reactions promoting decomposition may also have occurred. These processes affect both the yield and the integrity of the DNA sequences that are recovered. Nucleic acids from ancient specimens are often highly fragmented, and yields are typically reduced more than 1,000-fold relative to contemporary tissues (Rogan and Salvo 1991). Ancient nucleic acids are particularly susceptible to cleavage by endonucleases that excise oxidized pyrimidine moieties (Pääbo 1989). In some of the South American samples that we have analyzed, large genomic fragments greater than 10 kilobase pairs have been imaged by electron microscopy (Salvo unpubl. observations). However, interstrand crosslinking is also present and retards the migration of aDNA during agarose gel electrophoresis (Rogan and Salvo 1990; Salvo, unpubl. observations). We have previously shown that ancient nucleic acids can be enzymatically copied with certain DNA- or RNA-dependent DNA polymerases (AMV reverse transcriptase and Klenow DNA polymerase) that bypass some of these damaged sites (Rogan and Salvo 1990). The replicates can then serve as bonafide templates for PCR amplification with *Taq* DNA polymerase.

Extensive DNA damage and limited yields of aDNA impose substantial constraints on successful polymerase chain amplification that are not encountered when working with contemporary specimens. Modification of these templates presumably can introduce mutations into the resulting sequence and limit the length of the longest sequences that can be successfully amplified (Pääbo et al. 1990). The PCR products may therefore not be of sufficient length or accuracy to confirm their ancient human origin.

PCR-based procedures capable of verifying the authenticity and homogeneity of ancient human templates can be applied to control against potential contamination from human or other eukaryotic, contemporary or ancient nucleic acids. Postmortem decay of human bodies, the burial context, or residual parasitic infections can lead to recovery of human nucleic acids contaminated with DNA derived from other genomes (Royal and Clark 1960; Higuchi and Wilson 1984). Human nucleic acids may be distinguished from orthologous sequences found in other organisms by amplification of loci, such as the ribosomal RNA gene loci, that are relatively monomorphic within each species but exhibit interspecies variability. On the other hand, ethnic and geographic variation in contemporary humans has been extensively catalogued by studying mitochondrial DNA (mtDNA) inheritance (Vigilant et al. 1991). It is conceivable that the mtDNA haplotypes of ancient-specimens might be sufficiently distinctive that contamination with contemporary human nucleic acids would be easily detected. Ultimately, the analysis of genomic polymorphisms, when accomplished, will uniquely identify each mummified individual.

Pretreatment of aDNA substrates with Klenow DNA polymerase reverse transcriptase (RT) greatly facilitates the subsequent retrieval of DNA sequences from these preparations (Rogan and Salvo 1990). Since there was the possibility that the pretreatment and amplification process could be mutagenic, a highly conserved sequence was studied in the experiments described below. The identification of other species that might have contributed to the pool of nucleic acids is clearly an important problem, since sequence errors introduced during the amplification of certain human sequences could, in theory, produce sequences resembling those of nonhuman species.

We wished to determine whether ancient nucleic acids might be inaccurately replicated, resulting in amplification products with sequences divergent from the original template. Redundant, multicopy sequences were selected to increase the probability of successful amplification and to simplify the detection of mutations which might have arisen during amplification. The 18S, 5.8S, and 28S ribosomal DNA (rDNA) genes together compose a single multicistronic sequence, with a similar number of copies of each present in the human genome (Fig. 1). There are at least 160 uniform copies of the rDNA multigene family (Schmickel 1973), organized in long tandem arrays at several sites in

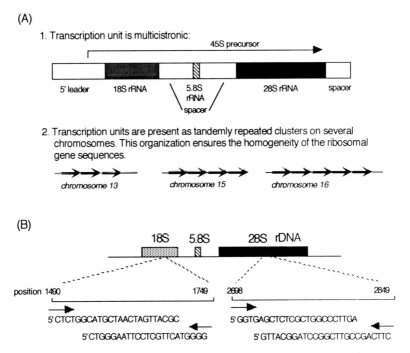

FIGURE 1. (A) Organization of large subunit rDNA genes in eukaryotes. The 18S and 28S rRNA genes are arranged on a multicistronic, tandemly repeated DNA segment in humans and other eukaryotes. Although several distinct clusters of rDNA repeats are found at different locations within the genome, the overall sequence relationships within the repeat unit are retained. The 18S and 28S rDNA sequences are conserved between species, although divergent sites may be used to establish taxonomic relationships (Rogan et al., 1993). (B) Locations of oligonucleotide PCR primers within the 18S and 28S rDNA genes. The PCR amplification products are defined by the corresponding coordinates in the 18S (GenBank locus HUMRGE) and 28S (GenBank locus HUMRGM) rDNA sequences.

the human genome (Henderson et al. 1972). These sequences are highly uniform, and the PCR products were predicted to have identical sequences unless the amplification reaction itself was mutagenic.

Any sequence heterogeneity found in the amplification products would be predicted to arise from chemical modification that has accumulated in the aDNA templates or from misincorporation during the PCR procedure. In order to ensure that human sequences were amplified, a 28S rDNA interval was selected which is known to exhibit a high degree of interspecies variability (Rogan et al., in press). The analysis of these loci therefore provides a means of assessing the overall integrity of the DNA, the accuracy of the amplification process, and potential contamination with other eukaryotic organisms.

2. Methods

2.1 DNA Isolation

Specimens were dissected with disposable scalpels and other autoclaved instruments. Only internal tissues were taken, to eliminate the possibility of contamination with contemporary DNA that could conceivably have arisen during excavation or subsequent handling of the mummy. Nucleic acids were purified from 2–11 g of skeletal muscle from eight different individuals based on a modification of the method described in Herrmann and Frischauf (1987). Samples were ground to a powder in liquid nitrogen with a Waring blender and digested overnight in a solution of proteinase K (100 μg/ml), sodium dodecyl sulfate (0.5%), 1 mM Tris-Cl (pH 8.0), and 100 mM EDTA at 37°C. The suspension was centifuged and the resulting supernatant was repeatedly extracted with a mixture of phenol and chloroform until the organic/aqueous interface was well defined. Nucleic acids were precipated from the aqueous fraction with ethanol or isopropanol, lyophilized, and dissolved in TE (10 mM Tris-Cl [pH 8.0], 1 mM EDTA) at the approximate concentration of 0.2 ml per gram of tissue. A reddish-brown contaminant that coprecipitated in some of the samples was removed by Sephadex G-50 spin column chromatography.

2.2 Choice Amplification Templates

The decision to select a reiterated, uniform sequence was in part based on previous results demonstrating the efficient amplification of such templates (Rogan and Salvo 1990). The 5' and 3' terminal coordinates of the 28S PCR product correspond to positions 2698 and 2849 of the published sequence (GenBank locus HUMRGM, accession number M11167; Gonzalez et al. 1985) while the 18S product spans nucleotides 1490 through 1749 (GenBank locus HUMRGE, accession number M10098; Torczynski et al. 1985). The predicted sizes of the 28S and 18S amplification products are 162 bp and 260 bp, respectively. The length of these amplified intervals does not exceed the limitations imposed by most aDNA templates. The left and right oligonucleotide primers derived from the 18S sequence contain natural *Sph*I and *Eco*RI restriction sites, which upon amplification were cleaved to generate cohesive termini suitable for subsequent cloning of the PCR-generated fragments.

One of the criteria used to select the rDNA primer sequences was their ability to amplify orthologous sequences from a wide variety of eukaryotic species. The sequences of rRNAs from different species can be aligned, and can be used to determine phylogenetic relationship between different organisms (Woese 1987). Differences between the aligned sequences form the basis for taxonomic classification schemes. The polymerase chain reaction has been used to amplify highly divergent sequence regions that are flanked by conserved domains corresponding to the oligonucleotide primers (Rogan et al., 1993).

The human 28S primer combination is capable of amplifying contemporary preparations of nucleic acids from a diverse set of plant, fungal, and animal species. The sequences of these amplification products have been used to generate a dendrogram illustrating the phylogenetic relationships between these organisms (Rogan et al. 1990).

2.3 Modified PCR Protocol

Experiments were carried out in a laboratory environment which, with the exception of control samples, was devoid of purified contemporary human genomic DNA. To prevent aerosol contamination between solutions, aDNA samples were manipulated with either a dedicated set of positive displacement pipettes or, when necessary, with conventional instruments equipped with cotton-plugged pipette tips.

In order to prepare intact genomic DNA templates for PCR amplification, replicates of aDNA were synthesized in an oligonucleotide-primed reverse transcriptase reaction. The ancient nucleic acid (5–10 ml) was denatured at 92°C for 5 min prior to annealing random hexanucleotide primers (1 mg), and addition of AMV reverse transcriptase (40 units). A 50 ml reaction containing 50 mM Tris-Cl (pH 8.0), 5 mM $MgCl_2$, 5 mM DTT, 50 mM KCl, 100 mM dNTPs, and 50 mg/ml BSA was incubated at 42°C for 2–3 hr. The reaction was stopped by adding 2 ml of 0.5 M EDTA. Oligonucleotide primers were then removed by Sephadex G-25 spin column chromatography, and the DNA was precipitated with ethanol, lyophilized, and resuspended in 25–40 ml of TE.

A 2–4 ml aliquot of pretreated aDNA was amplified by standard PCR for 40 cycles with *Taq* DNA polymerase (denaturation at 92°C, 1 min; annealing at 45°C, 1 min; extension at 72°C, 2 min; Saiki et al. 1985). Oligonucleotide primers were synthesized by Operon Technologies, Inc. (San Pablo, Cal.) and HPLC-purified prior to use. Primer sequences contained a GC-rich sequence at their 5′ terminus adjacent to a unique restriction site present in the rDNA sequence. The left and right 28S rDNA primers contained natural *Bam*HI and *Sac*I recognition sites, respectively, whereas *Eco*RI and *Sph*I sites were present in the 18S oligonucleotides. For each set of ancient specimens, both positive controls (contemporary DNA templates) and several negative controls (DNA template absent) were amplified in parallel.

One-fifth volume of each amplification reaction was analyzed by gel electrophoresis in a 4% gel consisting of a 3:1 mixture of Nusieve and molecular biology grade agarose. Gels were stained with ethidium bromide and the DNA amplification products were visualized by UV fluorescence.

2.4 DNA Cloning and Sequencing

For two different mummified individuals, 5 mg of the 18S amplification product was extracted with phenol/chloroform, reprecipitated, and doubly digested with an excess of *Eco*RI and *Sph*I (100 units). Enzymes were inactivated by heat treatment. After adjusting the sample volume to 1 ml, oligonucleotides and short PCR products that could interfere with the subsequent cloning steps were removed by centrifugation through a Centricon-100 column (Amicon). The nucleic acids were then concentrated by ethanol precipitation and resuspended at 0.2–0.3 mg/ml. Forced ligation between a fivefold excess of the restriction-digested PCR products and similarly digested M13mp19 vector DNA was carried out according to standard protocols (Sambrook et al. 1990). Ligation products were transformed into *E. coli* strain JM103 and plated in YT soft agar containing IPTG and XGal (Messing1983). Colorless plaques containing recombinant phage were picked and single-stranded DNA was purified from independent cultures. The Sanger sequencing procedure was carried out with a modified T7 DNA polymerse (Tabor and Richardson 1989). Sequences were compared with the contemporary 18S consensus using the GCG sequence analysis package (Devereux et al. 1984).

3. Results

3.1 Efficiency of aDNA Amplification

Most of the specimens selected were successfully amplified regardless of age or geographic location of the burials. The study reported here included two individuals derived from the coastal Camerones site approximately 50 km south of Arica, Chile (CAM8-T8 and CAM8-T10), three mummies from the Morro site in the city of Arica (Mol-6 T7, Mol-T28 C9, and Mol-T22 C6), and two from San Miguel de Azapa approximately 15 km inland (AZ141-T249M, AZ141-T249I, and AZ28-T6B). Carbon-14 dating studies have demonstrated that all of these specimens are most likely pre-Columbian, with age estimates of approximately 550 years for the Camerones specimens, 1,200 for those from Santa Cruz de Azapa, and 2,000 years for those from the Morro cemeteries (M. Allison, pers. comm.). The climate and natural mechanism of preservation appear to be the only constants among the different specimens.

For each sample, both 18S and 28S rDNA amplification products of the correct size were synthesized. Figure 2 shows the 28S amplification products from seven different individuals separated by agarosed gel electrophoresis (4% Nusieve agarose). Generation of these products required a pretreatment with oligonucleotide-directed RT, consistent with our previous results (Rogan and Salvo 1990). A control amplification reaction with 50 ng of contemporary human placental DNA template shows the expected 28S rDNA product (Fig. 2, lane Jy). Multiple control reaction in which distilled water was subsituted for the solution containing DNA at either the prerepair or the amplification step of the procedure did not generate 18S or 28S PCR products (for example, see Fig. 2, lane θ).

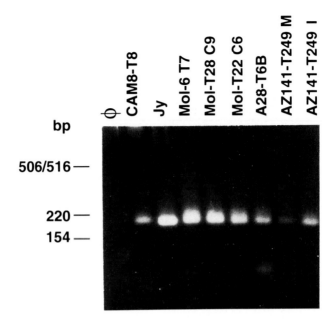

FIGURE 2. Amplification of a segment from human 28S rDNA. One specimen from the Camarones site (CAM8-T8), three from the Morro I cemetery (MoI-6 T7, MoI-T28 C9, and MoI-T22 C6), and three from San Miguel de Azapa (A28-T6B, AZ141-T249M, and AZ141-T249I) were amplified according to experimental conditions described in Rogan and Salvo (1990). The expected 162-bp product was generated by amplification of 50 ng of human placental genomic DNA (lane Jy). In the negative control reaction shown here (lane ⊖), 4 ml of double-distilled H$_2$O was substituted for template DNA at the AMV RT pretreatment step. The positions of size markers are shown in base pairs (bp).

3.2 Fidelity of aDNA Amplification

The accuracy of the amplified sequence is predicted to depend on both the integrity of the DNA template and the propagation of replication errors that could arise during the PCR process. This is because chemically modified sites can be misread by DNA polymerase during template replication (Clark and Beardsley 1987). It is conceivable that the PCR procedure could therefore incorporate mutations present on different replicated strands from one or more templates into the same amplification product.

To assess the fidelity of the PCR procedure, the sequences of individual, ancient 18S rDNA amplification products were compared with the contemporary rDNA sequence. The fragments were concentrated, restriction digested to generate cohesive ends and cloned into M13mp19 (Sambrook et al. 1990). Sequences of representative recombinant phage inserts were determined by Sanger dideoxy sequencing of seven different recombinant sequences

CAM 8 - T8   CAM 8 - T10

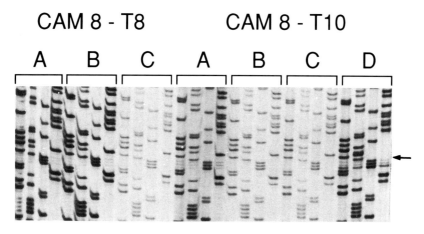

FIGURE 3. Sanger dideoxy sequencing the M13mp19 recombinants containing amplified 18S rDNA. 18S ribosomal RNA gene sequences were amplified from DNA isolated from two ancient specimens (CAM8-T8 and CAM8-T10), digested with *Eco*RI and *Sph*I, and cloned into M13mp19. Single-stranded DNA was prepared and sequences from seven clones were determined (three from CAM8-T8 denoted A, B, C, and four from CAM8-T10 denoted A, B, C, D). In each case, the order of sequencing reactions from left to right is TGCA. The arrow shows the position of the mutation in clone D from CAM8-T10. Sequencing reactions were carried out with Sequenase 2.0 (USB) and the M13 universal sequencing primer (GTTTTCCCAGTCACGAC) on 6% denaturing polyacrylamide gels. The immediate region around the discordancy is presented.

determined, four derived from amplification of CAM8-T10 and three from CAM8-T8 (Fig. 3).

The fidelity of an amplified sequence is predicted to be a function of the density and distribution of modified sites as well as the ability of RT and subsequently *Taq* DNA polymerase to bypass or correctly recognize modified positions. By analyzing the sequences of individual amplification products separately, it was possible to discern whether the amplification procedure was robust. Several different outcomes could be anticipated for this experiment. The presence of distinct mutations in each of the recombinant PCR products would suggest either that the amplification procedure is error-prone and/or that multiple, damaged templates were amplified. If most or all of the cloned sequences were similar to one another but differed from the reference sequence, it is likely that a limited number of damaged templates were amplified in a high fidelity PCR reaction. If the reference and cloned sequences were identical, either undamaged templates were present or RT presumably recognized the modified sites and incorporated the correct complementary nucleotide, effectively repairing the damage.

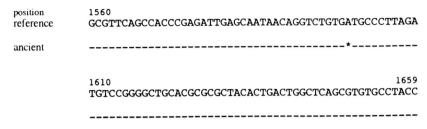

position   1560
reference   GCGTTCAGCCACCCGAGATTGAGCAATAACAGGTCTGTGATGCCCTTAGA

ancient   --*----------

                 1610   1659
                 TGTCCGGGGCTGCACGCGCGCTACACTGACTGGCTCAGCGTGTGCCTACC

FIGURE 4. Sequences of cloned PCR amplification products derived from aDNA and the contemporary 18S rDNA reference sequence (Gonzalez et al. 1985). Comparison of seven independent recombinants with the reference sequence revealed one nucleotide discordancy in a single clone (a transition indicated with an asterisk at position 1599). All other sequences were identical (–) to the contemporary sequence. Although the complete insert was sequenced for each recombinant, only the interval between nucleotides 1560 and 1659 of the reference sequence is shown.

Except for a single-nucleotide discordancy in one of the clones (a transition at position 1599 of the 18S sequence), each of the sequences determined was identical to the contemporary reference sequence (Fig. 4). There is no evidence that these samples were contaminated by nucleic acids from any other eukaryotic species. Several interpretations of this result are possible. Either unmodified templates were present in these samples, or the pretreatment and amplification enzymes are capable of correctly reading damaged nucleotides and adding the appropriate base at those sites, or a full-length template was reconstructed from intact segments derived from different rDNA templates during the amplification reaction. The single-nucleotide mismatch that was detected could be consistent either with modification of the templates or with the introduction of mutations during the PCR procedure (see below).

4. Discussion

A multicopy homogeneous template derived from the 18S-28S rDNA transcription unit was successfully amplified with a two step replication reaction, utilizing AMV RT followed by thermal cycling with TAQ DNA polymerase. The sequences of seven independent recombinant 18S reaction products from two different individuals revealed one nucleotide discordancy in over 1,000 nucleotides. This high level of fidelity is comparable to that found in conventional PCR amplification and may be due to the multicopy nature of the template, pretreatment with AMV-RT, or a combination of both of these factors.

4.1 High Fidelity of Amplification

During the amplification of aDNA sequences in vitro, chemically modified nucleotides may be sometimes misread by DNA polymerase, resulting in mutations in the amplified sequence (Pääbo et al. 1990). In the present study, the accuracy of a modified PCR protocol was assessed for two different highly conserved, multicopy genomic templates using ancient human DNA preparations from eight distinct individuals. Sequence analysis shows that both 18S and 28S rDNA products are generated in each of the DNA samples and that 18S sequences are accurately amplified. This suggests either the presence of intact genomic templates that were replicated with high fidelity or that the hexanucleotide-directed RT was capable of recognizing modified nucleotides and incorporating the appropriate complementary bases at these sites.

A single-nucleotide discordancy relative to the contemporary consensus was found in the sequence of one of the cloned PCR products. The error may have been introduced during the later stages of amplification rather than the prerepair step, since mutations arising during the RT reaction would be propagated throughout the remaining cycles, resulting in identical changes in multiple recombinant clones. This mutation represents approximately 0.1% of the total sequence determined. Since *Taq* DNA polymerase misincorporates approximately 1 in 104 nucleotides in vitro, the observed sequence error could have occurred during amplification (Tindall and Kunkel 1988). The difference between the observed and the experimentally determined enzymatic mutation rate is probably not statistically significant.

We cannot completely eliminate the possibility that the observed sequence change arose during replication of the aDNA template with RT. From these experiments, it is not possible to determine which strand was mutated, only that a A → G or a T → C transition occurred. RT may have misread and/or misincorporated at this site as thymidine is a predominant target for oxidative damage to DNA (Hutchinson 1985). Pyrimidine substitutions represent the majority of oxidative damage to DNA (Hutchinson 1985). They also represented the majority of mutations produced in the amplification of ancient Egyptian mtDNA sequences (Pääbo et al. 1990).

4.2 High Efficiency of Amplification

Our current and previous efforts demonstrate that human repetitive DNA sequences can be amplified in many of the pre-Columbian, mummified specimens that have been analyzed (Rogan and Salvo 1990). Successful amplification at these loci may be due, in part, to the multicopy stoichiometry of the template. Although the copy number of rDNA templates is at least 5,000-fold lower than that of the *Alu* elements previously studied, this sequence is more highly conserved. The similar copy number of rDNA and mtDNA in the cell may explain why ancient mtDNA sequences can be consistently retrieved (Pääbo et al. 1988; Rogan and Salvo 1990). The inability to amplify these sequences in some samples by conventional approaches suggests that most of the templates in these preparations may be chemically modified. In each repetitive sequence family there may be a small number of nearly intact copies, which are preferentially amplified when treated with a DNA polymerase that bypasses the damaged sites.

These DNA lesions appear to partially inhibit the PCR amplification reaction by blocking extension by some DNA polymerases (Clark and Beardsley 1987). A full-length amplification template is thought to be generated by the repeated annealing and replication of overlapping, complementary partial extension products (Pääbo et al. 1990). The kinetics of the amplification reaction are predicted to be biphasic: exponential amplification would not be expected to occur until the full-length product was synthesized. The formation of a detectable PCR product would be predicted to require additional cycles of amplification.

DNA synthesis using oligonucleotide-directed RT prior to amplification may provide a machanism for bypassing some of the chemically modified sites present in aDNA. Annealing of hexanucleotides composed of random sequences permits AMV RT to intiate DNA synthesis at many different priming sites along the single-stranded aDNA templates. RT has a propensity to generate incomplete extension products from RNA or DNA templates both during retroviral replication and in vitro. Both Klenow DNA polymerase and RT recognize and replicate beyond sites of oxidative damage in vitro, although this reaction is inefficient (Clark and Beardsley 1987). This might result in longer partial extension products from aDNA templates than those generated by *Taq* DNA polymerase. The increased length of these intermediate species should facilitate the formation of complementary duplexes capable of serving as full-length templates. Consistent with this prediction, major extension products can be detected in the early stages of PCR amplification using ancient templates pretreated with RT. These intermediates appear to be converted into full-length products with retarded kinetics compared to contemporary DNA controls, and are not present in the control reactions.

Although these studies were designed to utilize the DNA-dependent DNA polymerase activity of RT, it is possible that preserved ribosomal RNA species may have been amplified. The protocol for isolating ancient nucleic acids does not degrade RNA by treatment with ribonuclease A. 18S and 28S rRNAs are among the most abundant nucleic acid species in living cells, and it is conceivable that sufficient template may have remained to generate DNA products from a random-primed reverse transcription reaction. Venanzi and Rollo (1990; cf. Rollo et al. Chapter 16) have presented evidence that ancient nucleic acids from plant specimens are composed mainly of RNA. The apparent preservation of RNA in ancient preparations is remarkable, given the seeming ubiquity of ribonucleases that contribute to RNA lability in vitro. The integrity of rRNA could have conceivably been protected from nuclease action either by endogenous or postmortem chemical modification or by interactions with preserved ribosomal proteins. Future studies will address this issue.

Acknowledgments. We would like to thank Drs. Marvin Allison and Arthur Aufderheide for generously providing ancient specimens and their continued support and encouragement of this research. P.K.R. is grateful to Dr. Paul Tooley of the USDA Plant Pathology Laboratory for providing laboratory facilities in which the PCR amplification studies were carried out.

References

Clark J, Beardsley GP (1987) Functional effects of *cis*-thymine glycol lesions on DNA synthesis in vitro. Biochemistry 26:5398–5403

Devereux J, Haeberli, Smithies O (1984). A comprehensive set of sequence analysis programs for the VAX. Nucl Acids Res 12:387–395

Gonzalez IL, Gorski JL, Campen TJ, Dorney DJ, Erickson JM, Sylvester JE, Schmickel RD (1985) Variation among human 28S ribosomal RNA genes. Proc Natl Acad Sci USA 82:7666–7670

Henderson AS, Warburton D, Atwood KC (1972) Location of ribosomal DNA in the human chromosome complement. Proc Natl Acad Sci USA 69:3394–3398

Herrmann BG, Frischauf AM (1987) Isolation of genomic DNA. Meth Enzymol 152:180–183

Higuchi R, Wilson A (1984) Recovery of DNA from extinct species. Fed Proc 43:1557

Hutchinson F (1985) Chemical changes induced in DNA by ionizing radiation. Prog Nucleic Acid Res Mol Biol 32:115–154

Messing J (1983) New M13 vectors for cloning. Meth Enzymol 101:20–78

Pääbo S (1989) Ancient DNA: extraction, characterization, molecular cloning, and enzymatic amplification. Proc Natl Acad Sci USA 86:6196–6200

Pääbo S, Gifford JA, Wilson AC (1988) Mitochondrial DNA sequences from a 7,000-year old brain. Nucl Acids Res 16:9775–9787

Pääbo S, Irwin DM, Wilson AC (1990) DNA damage promotes jumping between templates during enzymatic amplification. J Biol Chem 265:4718–4721

Rogan P, Salvo J (1990) Molecular genetics of pre-Columbian South American mummies. UCLA Symposium on Molecular Evolution 122:223–243

Rogan P, Salvo J (1991) Study of nucleic acids isolated from ancient remains. Yrbk Phys Anthrop 33:195–214

Rogan P, Salvo J, Tooley P (1990) Use of universal PCR primers amplify 28S ribosomal DNA from taxonomically diverse organisms. Proceedings of the 4th International Congress of Systematic and Evolutionary Biology, vol. 2. College Park, Md: University of Maryland, p. 393

Rogan P, Salvo J, Stephens RM, Schneider T (1993) Design of universal polymerase-chain reaction primers for amplification of 28S rDNA. In: Pickover C (ed) *The Visual Display of Biological Information*, World Scientific, River Edge, NJ.

Royal W. Clark E (1960) Natural preservation of human brain, Warm Mineral Springs, Florida. Am. Antiquity 26:285–287

Saiki R, Gelfand D, Storfei S, Scharf S, Higuchi R, Horn G, Mullis K, Erlich H (1988) Primer-directed enzymatic amplification of DNA with a thermostable DNA polymerase. Science 239:487–491

Sambrook J, Fritsch E, Maniatis T (1990) *Molecular Cloning: A Laboratory Manual*, 2nd ed. Cold Spring Harbor, N.Y.: CSHL Press

Schmickel RD (1973) Quantitation of human ribosomal DNA: hybridization of human DNA with ribosomal RNA for quantitation and fractionation. Pediat Res 7:5–12

Tabor S, Richardson CC (1989) Selective inactivation of the exonyclease activity of bacteriophage T7 DNA polymerase by in vitro mutagenesis. J Biol Chem 214:6447–6458

Tindall K, Kunkel T (1988) Fidelity of DNA synthesis by the *Thermus aquaticus* DNA polymerase. Biochemistry 27:6008–6013

Torczynski RM, Fuke M, Bollon AP (1985) Cloning and sequencing of a human 18S ribosomal RNA gene. DNA 4:283–291

Venanzi F, Rollo F (1990) Mummy RNA lasts longer. Nature 343:310–311

Vigilant L, Stoneking M, Harpending H, Hawkes K, Wilson AC (1991) African populations and the evolution of human mitochondrial DNA. Science 253:1503–1507

Woese C (1987) Bacterial evolution. Microbiol Rev 51:221–271

Dried Samples: Hard Tissues
13 Mitochondrial DNA from Ancient Bones

ERIKA HAGELBERG

1. Introduction

Most research on ancient DNA (aDNA) has involved the cloning or amplification of DNA extracted from soft tissue remains (Higuchi et al. 1984; Pääbo 1985; Pääbo et al. 1988; Thomas et al. 1989; Golenberg et al. 1990; Lawlor et al. 1991). The discovery that DNA can also be recovered from ancient bones (Hagelberg et al. 1989; Horai et al. 1989; Hänni et al. 1990; Hummel and Herrmann 1991) has created new possibilities for the study of past populations, as bones are abundant archaeological remains and many museums throughout the world contain extensive and well-characterized osteological collections.

As with other aDNA studies (Pääbo et al. 1988; Pääbo 1989; Pääbo et al. 1989), research on DNA from bone has concentrated on the analysis of mitochondrial DNA (mtDNA) fragments. This is because mtDNA is well characterized and the high copy number of mtDNA in animal cells favors its survival in decayed tissues, while its fast rate of evolution and maternal mode of inheritance are useful for evolutionary and phylogenetic studies. The polymerase chain reaction (PCR) is the ideal technique for the analysis of aDNA, as most specimens can be expected to contain very little, frequently damaged DNA.

Recently we showed that it is relatively straightforward to amplify long mtDNA fragments (up to 1 kb) from ancient bones (Hagelberg and Clegg 1991; Hagelberg et al. 1991a); indeed it is possible that DNA may survive better in bone than in soft tissue remains (cf. Cooper, Chapter 10) as in general, longer fragments can be amplified from bone. Bone DNA typing by PCR has also been shown to be sufficiently reliable for forensic investigations, as demonstrated in two recent studies where human skeletal remains were identified, in one case by PCR and sequencing of mtDNA (Stoneking et al. 1991) and in the other by amplification of polymorphic nuclear DNA microsatellites (Hagelberg et al. 1991b).

Despite these developments and the opportunities offered by bone DNA typing for anthropology, molecular evolution, population genetics, and forensic science, this work is still associated with a number of difficulties and potential pitfalls. Ancient remains including bone can become contaminated by human DNA from shed human skin flakes during excavation and conservation. Ancient specimens are additionally contaminated with bacteria, fungi, and insects, as a result both of natural decay processes in the soil and of storage in museums and collections; the nucleic acids from these organisms are coextracted with those of the specimens and constitute the major proportion of the DNA in the tissue extracts. Trace amounts of human DNA in laboratory reagents present a further contamination hazard to be avoided at all costs, particularly when ancient human remains are analyzed, as it may be impossible to distinguish between the original DNA from a specimen and exogenous contamination.

Whether research involves human or animal bones, the problems of contamination begin in earnest once work starts on the amplification of the ancient extracts, because millions of copies of the DNA segments of interest are produced in each reaction. Stringent procedures should be followed to avoid contamination of the whole laboratory environment with these DNA fragments. Some of the procedures we use to deal with problems of contamination are described in this chapter, as are experimental protocols for the extraction, amplification, and sequencing of DNA from ancient bone specimens.

2. Methods

2.1 General Considerations

Most aDNA research takes place within larger molecular biology laboratories, and it is important to work as carefully as possible to avoid contamination by modern DNA. Disposable sterile containers and pipette tips, and sterile reagents and solutions dedicated for work on aDNA should be used. Bone specimens should *never* be blown on to remove soil or dust, and breathing and talking over DNA work should be avoided.

Working solutions can be prepared with bottled sterilized water available in sealed 1-liter plastic bottles from medical suppliers, this is cheap and convenient, and the empty bottles provide excellent (and sterile) containers for buffers. The solutions are made up quickly, with minimum exposure to the ambient air; pH electrodes are never placed in the bulk solution, but rather a small amount of the solution poured into a separate container and tested. For most solutions, small inaccuracies in pH are less important than the contamination hazard. The cleanliness of the reagents and containers should be checked by performing blank control extractions (containing no bone) in parallel with the bone extractions.

2.2 Specimen Selection and Preparation

There is significant variability in the quality of bone specimens available, and the survival of DNA in an individual bone depends on its state of preservation rather than age. We have found profound differences on occasion in the amplification of DNA from bone samples of the same age and recovered from the same archaeological site, presumably caused by local differences in the burial environment. Generally, however, we observe a reasonably good correlation between external gross preservation and the ability of a specimen to yield amplifiable DNA (Hagelberg et al. 1991a).

In optimal circumstances, specimens should be freshly excavated and unwashed. Naturally this will not always be possible; but bones that have been washed and stored damp in plastic bags can become moldy and should be avoided. If there is a choice, we prefer to select a fragment of cortical long bone like tibia or femur; a wedge weighing a few grams can be removed with a hacksaw, with minimum damage to the specimen. We usually find, however, that we need to work with a wide range of different samples, both cranial and postcranial, because they are the only ones available for a particular study. When human bone specimens are analyzed, it is advisable to include some animal bones from the same archaeological context as an additional contamination control; unambiguous sequences from the correct animal species increase the confidence of results obtained from the human bones processed in parallel.

Soil or clay adhering to the specimens can be cleaned off with a scalpel blade. It is advisable to remove as much material as possible from the outside surface of the bones, in case it is contaminated. This can be achieved by sandblasting the bone pieces with fresh abrasive grit; we use an Airbrasive 6500 unit (S.S. White Industrial Products, Piscataway, N.J.), although satisfactory results can be achieved with fine emery paper or simply by scraping the bones well with a clean scalpel blade. If a sandblaster is used, the abrasive grit (alumina or carborundum) should not be recycled.

After cleaning, the specimens are broken into small fragments (0.5–1 cm) with a hammer or pliers and ground up in a cryogenic impact grinder with self-contained liquid nitrogen tub (Spex Industries Inc., Edison, N.J.), using grinding vials with the capacity to hold 2–4 g bone. The grinding vials and impactors of the mill are cleaned and sterilized after use to prevent sample carryover. After grinding, the samples of bone powder are stored in sterile labeled containers at $-20°C$.

2.3 DNA Extraction

The extractions are usually done with 1 g powdered bone, and this amount may produce enough DNA for 100–200 amplifications. However, there is no reason not to scale down the extractions and alter the conditions to suit the state of preservation of individual specimens. We describe here a typical extraction with 1 g bone powder.

The powder is weighed on a disposable weighing boat or piece of aluminum foil and placed in a labeled, sterile 50-ml polycarbonate tube with a screw cap. Extensive decalcification with EDTA is unnecessary, but the powder can be washed two or three times by centrifugation with 10–15 ml of an EDTA solution (the concentration is immaterial: we use the same 0.5 M EDTA, pH 8–8.5 solution employed in the lysis buffer) to remove the soluble brown impurities. After these washing steps, 10 ml of lysis buffer is added to each sample tube. The bone lysis buffer is composed of 0.5 M EDTA, pH 8–8.5, 200 µg/ml proteinase K, and 0.5% N-lauroylsarcosine (adapted from Maniatis et al. 1982).

The caps of the tubes should be firmly tightened to avoid leaks. The tubes (labeled with adhesive tape) can be incubated either at 37°C overnight or at 50°C for 3 hr, with thorough agitation to prevent the bone powder from settling. We put the sample lysis tubes horizontally in a plastic box, with a tight-fitting lid on a shaker in a hot room or shaking water bath; this way, the outside of the tubes remains clean.

Solid matter remaining in the lysis suspension after the incubation step will be subsequently removed by centrifugation during the phenol/chloroform extraction steps, which are as follows:

- One extraction with an equal volume of phenol (freshly opened water-saturated phenol, glass distilled grade, Rathburn Chemicals Limited, Wakeburn, Scotland; equilibrated with 20 mM Tris-HCl, pH 7.5). The aqueous phase remains at the bottom (provided the lysis buffer contained 0.5 M EDTA), and the top phenol layer is discarded.
- Two extractions with an equal volume of phenol/chloroform (1:1). The aqueous phase is at the top, and the organic phase and solid matter remain at the bottom.
- One or two extractions with chloroform to remove all traces of phenol.

The aqueous phase is desalted and concentrated by centrifugation dialysis at room temperature, using Centricon-30 microconcentrators (Amicon, Danvers, Mass.) in a benchtop or floorstanding centrifuge according to the manufacturer's advice. As the capacity of the microconcentrators is only 2 ml, several centrifugation steps are required to concentrate all the aqueous solution. Finally, the retentate (100–200 µl) is washed at least twice with 2 ml sterile water. The resulting bone extract contains DNA (both the original bone DNA and the nucleic acids from microorganisms in the bones) as well as carbohydrates, some protein, and impurities like humic acids. Aliquots (5–10 µl) of this DNA are electrophoresed through 1% agarose gels with DNA size markers, and the gels are then stained with ethidium bromide to visualize the DNA under UV light. Like other workers, we sometimes see a blue fluorescence when the ethidium bromide gels are examined on a UV transilluminator, presumably caused by soil-derived humic acids in the bone DNA extracts. The humic acids are a heterogeneous mixture of large aromatic molecules with many reactive carboxyl and hydroxyl side groups and a distinctive pattern of fluorescence, which is presumably what is observed on the agarose gels under UV light.

High molecular weight DNA is frequently observed on the electrophoresis gels of bone extracts, but most of it stems from microorganisms. Analysis of bone extracts by Southern blot hybridization with a human *Alu* repeat probe has shown that only a small proportion of the extracted DNA is of human origin, and this DNA is severely degraded to fragments smaller than a few hundred nucleotides (Hagelberg et al. 1991b; cf. Nielsen et al., Chapter 8).

2.4 DNA Amplification

We purchase as many ready-made solutions as possible, rather than risking contamination of the reagents with DNA from our laboratory, and find the additional expense small compared with the convenience and improved quality of the results. A commercial 10X PCR buffer, 25 mM $MgCl_2$, and nucleotides can be purchased from Perkin Elmer Cetus. The water used for PCR is sterile water for injection, available in sealed 10-ml ampoules or 50-ml bottles from medical suppliers. Once opened, solutions are aliquoted into sterile microfuge tubes, labeled with the contents and date, and stored at $-20°C$. The use of aliquoted reagents and primers protects bulk solutions from contamination.

Amplifications are carried out by the method recommended by Perkin Elmer Cetus in 25-µl reactions. It is advisable to set up a master mix with all the reagents necessary for the number of reactions in a given experiment (with the exception of the template DNA) to avoid the inaccuracy inherent in pipetting small volumes, then mix this solution thoroughly as the *Taq* polymerase contains glycerol and will settle out otherwise, and finally to aliquot the necessary volume into the separate reaction tubes. The template DNA is then added to the reaction buffer in each tube, and the whole is overlaid with a drop of sterile mineral oil. The final reaction buffer consists of 10 mM Tris-HCl, pH 8.3, 50 mM KCl, 2.5 mM $MgCl_2$, 0.01% gelatin, 200 µM of each dNTP, 20 pM of each primer, 160 µg/ml bovine serum albumin, and 2 units (*Thermus aquaticus*) *Taq* DNA polymerase. Typically 1–1.5 µl of bone DNA extract is used in each amplification reaction; each experiment also includes two control reactions set up in parallel, one a blank reaction with no DNA to check the purity of the PCR reagents, and the second an extraction control to check the purity of the DNA extractions.

Typical PCR conditions for mtDNA fragments are 35 cycles of amplification, with denaturation at 94°C (1 min), annealing at 55°C (1 min), and extension at 72°C (1 min); the denaturation step of the first cycle is increased to 5 min to ensure complete denaturation of the genomic DNA. Some bone extracts may require two rounds of PCR (Horai et al. 1989) to give a positive amplification result; this involves taking a small aliquot of the PCR product after 30–35 cycles of PCR and adding it to fresh reaction buffer, which is amplified again for 30–35 cycles. We would advise extreme caution in the interpretation of results obtained from such a high number of PCR cycles, because of the increased danger of amplification of trace amounts of contaminating DNA.

Aliquots (5–10 μl) of the PCR reaction are analyzed by electrophoresis on agarose gels stained with ethidium bromide to visualize the DNA fragments under UV light, or alternatively on polyacrylamide gels stained with ethidium bromide or silver stain (Bio-Rad Silver Stain, Bio-Rad, Richmond, Cal.). Particular care should be taken when handling PCR products to prevent carryover contamination.

Table 1 lists some of the primers that have been found to be useful for amplifying mtDNA fragments from ancient bone. Good results with human and animal bones are obtained with the highly conserved mtDNA primers L14841 and H15149 which amplify a 375-base-pair (bp) fragment of the cytochrome *b* gene, and primers L1091 and H1478 which amplify a 449-bp fragment of the 12S rRNA gene (Kocher et al. 1989). These primers amplify homologous sequences in a wide range of animal species and have been used for phylogenetic studies (Kocher et al. 1989; Thomas et al. 1989, 1990; Irwin et al. 1991). A truncated version of the cytochrome *b* primers has also been described (Kocher and White 1989).

Other useful primer pairs are L15996/H16401 and L29/H408, which each amplify approximately 400 bp of the hypervariable region of human mtDNA (Vigilant et al. 1989). Primer H16401 can be used with D18 to amplify a 228-bp fragment of the hypervariable region (Higuchi et al. 1988). Primers A, B, and C are in a small noncoding region (region V) of mtDNA. Primers A and B amplify a 120-bp fragment (primers A and C a 91-bp fragment) containing a useful anthropological marker for individuals of Asian origin (Wrischnik et al. 1987).

Like other workers (Pääbo et al. 1988; Pääbo 1990), we find that many aDNA extracts contain a powerful PCR inhibitor. In most cases inhibition can be overcome by adding bovine serum albumin to the amplification buffer, and by diluting the extract and so diluting the inhibitor. Other workers have successfully used methods involving spermidine and chelating resins (cf. e.g. Ellegren, Chapter 15; Taylor & Swann, Chapter 11).

2.5 Asymmetric PCR and DNA Sequencing

The amplification products can be sequenced directly using a modified unbalanced PCR method (Gyllensten and Erlich 1988; Hagelberg and Clegg 1991) to generate a single-stranded product suitable for sequencing. First, a normal PCR amplifications of the bone extract is performed using equal concentrations of both primers. A second amplification is then done with 1 μl of the first reaction as the template and omitting one of the two primers.

TABLE 1. Some primers used for the amplification of mtDNA fragments from ancient bone. Base numbers refer to Anderson et al. (1981).

Primer	Sequence	Size (bp)	Reference
Region V			
A	5'-ATGCTAAGTTAGCTTTACAG-3'-8297	20	Wrischnick et al. 1987
B	5'-ACAGTTTCATGCCCATCGTC-3'-8215	20	"
C	5'-ATTCCCCTAAAAATCTTTGA-3'-8244	20	"
Control Region			
L15996	5'-CTCCACCATTAGCACCCAAAGC-3'-15996	22	Vigilant et al. 1989
H16401	5'-TGATTTCACGGAGGATGGTG-3'-16401	20	"
L29	5'-GGTCTATCACCCTATTAACCAC-3'-29	22	"
H408	5'-CTGTTAAAGTGCATACCGCCA-3'-408	22	"
D18	5'-CCATGCTTACAAGCAAGT-3'-16219	18	Higuchi et al. 1988
Cytochrome b			
L14841	5'-AAAAAGCTTCCATCCAACATCTCAGCATGATGAAA-3'-14841	35	Kocher et al. 1989
H15149	5'-AAACTGCAGCCCCTCAGAATGATATTTGTCCTCA-3'-15149	34	"
12S rRNA			
L1091	5'-AAAAAGCTTCAAACTGGGATTAGATACCCCACTAT-3'-1091	35	Kocher et al. 1989
H1478	5'-TGACTGCAGAGGGTGACGGGCGGTGT-3'-1478-3'-1478	28	"

The limiting primer (carried over from the first PCR reaction) is used up after several PCR cycles, and the reaction proceeds using the second primer, resulting in an excess of single-stranded PCR product suitable for sequencing. The second round of PCR can be carried out in duplicate 100-µl reactions to increase product yield. The two reactions are pooled and the PCR products purified by centrifugation through Miniprep Spun Columns of Sephacryl S-400 (Pharmacia LKB Biotechnology AB, Uppsala, Sweden). The column eluate is ethanol precipitated and resuspended in a small volume of water, say 20 µl, and sequencing performed using 7 µl of this DNA and the limiting primer from the second amplification. A commercial kit (Sequenase; US Biochemical Corp.) is used for dideoxy chain-termination sequencing (Sanger et al. 1977), and the products are electrophoresed on 6% denaturing poly-acrylamide gels and autoradiographed for 2–3 days. Fairly high levels of background are observed on the sequencing autoradiographs, probably due to the method used to prepare the templates.

3. Discussion

The ability to retrieve genetic information from ancient bones has many uses in population and evolutionary studies. Most of the studies to date have concentrated on the techniques themselves rather than in answering particular research questions, but the unforeseen success of bone DNA typing in the forensic identification of skeletal remains (Hagelberg et al. 1991b) as well as the consistent results obtained with many archaeological specimens have increased our expectations for the potential of these techniques. Despite the methodological difficulties we envisage may exciting applications in coming years.

In the last two years we have analyzed mtDNA polymorphisms in DNA recovered from skeletal remains from prehistoric Pacific islanders (Hagelberg and Clegg 1993). Some of the specimens in our study were excavated from sites associated with the Lapita cultural complex; it is generally believed that it was the decendants of the Lapita people who were eventually responsible for the colonization of Polynesia (Bellwood 1989). A significant amount of genetic work has been done on the present inhabitants of this area, but additional information gained from archaeological bone specimens will help overcome some of the distortions generated by events such as genetic drift and population bottlenecks over the last 3,000 years or so.

We have also started to analyze skeletal samples recovered from Khok Phanom Di in central Thailand, an unusually rich site occupied between 2000 and 1500 B.C. (Higham 1989; Higham and Bannanurag 1990). Analysis of the mtDNA control region of individuals buried at this site will help in the reconstruction of putative genealogies inferred from the organisation of the burials and shared skeletal abnormalities.

Although it is possible that technical difficulties and the degradation of the DNA in ancient bones may limit most future studies to the last few ten thousand years, this time span encompasses sufficient significant events in the history of humankind to leave room for much interesting research, notably on the peopling of the Pacific and the Americas, the introduction of agriculture, animal domestication, and the spread of infectious diseases. In the meantime, we remain optimistic that in a few exceptional circumstances DNA may well have survived for hundreds of thousands or even millions of years.

Acknowledgments. I would like to thank John Clegg and my other colleagues in the MRC Molecular Haematology Unit for helpful discussions during the development of this work. I am grateful to Jerry Hagelberg for his helpful comments on the manuscript. This project was supported by a research grant under the NERC Special Topic in Biomolecular Palaeontology.

References

Anderson S, Bankier AT, Barrell BG, de Bruijn MHL, Coulson AR, Drouin J, Eperon IC, Nierlich DP, Roe, BA, Sanger F, Schreier PH, Smith AJH, Staden R, Young IG (1981) Sequence organisation of the human mitochondrial genome. Nature 290:457–465

Bellwood PS (1989) The colonization of the Pacific: some current hypotheses. In: Hill AVS, Serjeantson SW (eds) *The Colonization of the Pacific: A Genetic Trail,* Oxford: Oxford University Press, pp. 1–59

Golenberg EM, Giannasi DE, Clegg MT, Smiley CJ, Durbin M, Henderson D, Zurawski G (1990) Chloroplast DNA sequence from a Miocene *Magnolia* species. Nature 344:656–658

Gyllensten UB, Erlich HA (1988) Generation of single-stranded DNA by the polymerase chain reaction and its application to direct sequencing of the HLA-DQA locus. Proc Natl Acad Sci USA 85:7652–7656

Hagelberg E, Clegg JB (1991) Isolation and characterisation of DNA from archaeological bone. Proc R Soc Lond B 244:45–50

Hagelberg E, Clegg JB (1993) Genetic polymorphisms in prehistoric pacific islanders determined by analysis of ancient bone DNA. Proc Roy Soc Lond B 252:163–170

Hagelberg E, Sykes B, Hedges R (1989) Ancient bone DNA amplified. Nature 342:485

Hagelberg E, Bell LS, Allen T, Boyde A, Jones SJ, Clegg JB (1991a) Ancient bone DNA: techniques and applications. Phil Trans R Soc Lond B 333:399–407

Hagelberg E, Gray IC, Jeffreys AC (1991b) Identification of the skeletal remains of a murder victim by DNA analysis. Nature 352:427–429

Hänni C, Laudet V, Sakka M; Begue A, Stehelin D (1990) Amplification of mitochondrial DNA fragments from ancient human teeth and bones. C R Acad Sci Paris 310 Series 3:365–370

Higham CFW (1989) The prehistoric site at Khok Phanom Di. Arts Asiatique 44(2):5–43

Higham CFW, Bannanurag R (1990) *The Excavation at Khok Phanom Di, a Prehistoric Site in Central Thailand.* Vol. 1, *The Excavation, Chronology and Human Burials.*

Research Report of the Society of Antiquaries of London, no. 47. London: Society of Antiquaries

Higuchi R, Bowman B, Freiberger M, Ryder OA, Wilson AC (1984) DNA sequences from the quagga, an extinct member of the horse family. Nature 312:282–284

Higuchi R, von Beroldingen CH, Sensabaugh GF, Erlich HA (1988) DNA typing from single hairs. Nature 332:543–546

Horai S, Hayasaka K, Murayama K, Wate N, Koike H, Nakai N (1989) DNA amplification from ancient human skeletal remains and their sequence analysis. Proc Jap Acad 65:229–233

Hummel S, Herrmann B (1991) Y-chromosome-specific DNA amplified in ancient human bone. Naturwissenschaften 78:266–267

Inwin DM, Kocher TD, Wilson AC (1991) Evolution of the cytochrome *b* gene of mammals. J Mol Evol 32:128–144

Kocher TD, White TJ (1989) Evolutionary analysis via PCR. In: Erlich HA (ed) *PCR Technology: Principles and Applications for DNA Amplification.* New York: Stockton Press, pp. 137–147

Kocher TD, Thomas WK, Meyer A, Edwards SV, Pääbo S, Villablanca FX, Wilson AC (1989) Dynamics of mitochondrial DNA evolution in animals: amplification and sequencing with conserved primers. Proc Natl Acad Sci USA 86:6196–6200

Lawlor AL, Dickel CD, Hauswirth WW, Parham P (1991) Ancient HLA genes from 7,500-year-old archaeological remains. Nature 349:785–788

Maniatis T, Frisch EF, Sambrook J (1982) Molecular Cloning: A Laboratory Manual. Cold Spring Harbor, N.Y.: CSHL Press

Pääbo S (1985) Molecular cloning of ancient Egyptian mummy DNA. Nature 314:644–645

Pääbo S (1989) Ancient DNA: extraction, characterization, molecular cloning, and enzymatic amplification. Proc Natl Acad Sci USA 83:1939–1943

Pääbo S (1990) Amplifying ancient DNA. In: Innis MA, Gelfand DH, Sninsky JJ, White TJ (eds) *PCR Protocols: A Guide to Methods and Applications.* San Diego: Academic Press, pp. 159–166

Pääbo S, Gifford JA, Wilson AC (1988) Mitochondrial DNA sequences from a 7,000-year-old brain. Nucl Acids Res 16:9775–9787

Pääbo S, Higuchi RG, Wilson AC (1989) Ancient DNA and the polymerase chain reaction. J Biol Chem 264:9709–9712

Sanger F, Nicklen S, Coulson AR (1977) DNA sequencing with chain-terminating inhibitors. Proc Natl Acad Sci USA 74:5463–5467

Stoneking M, Hedgecock D, Higuchi RG, Vigilant L, Erlich HA (1991) Population variation of human mtDNA control region sequences detected by enzymatic amplification and sequence-specific oligonucleotide probes. Am J Hum Genet 48:370–382

Thomas RH, Schaffner W, Wilson AC, Pääbo S (1989) DNA phylogeny of the extinct marsupial wolf. Nature 340:465–467

Thomas WK, Pääbo S, Villablanca FX, Wilson AC (1990) Spatial and temporal continuity of kangaroo rat populations shown by sequencing mitochondrial DNA from museum specimens. J Mol Evol 31:101–112

Vigilant L, Pennington R, Harpending H, Kocher TD, Wilson AC (1989) Mitochondrial DNA sequences in single hairs from a southern African population. Proc Natl Acad Sci USA 86:9350–9354

Wrischnik LA, Higuchi RG, Stoneking M, Erlich HA, Arnheim N, Wilson AC (1987) Length mutations in human mitochondrial DNA: direct sequencing of enzymatically amplified DNA. Nucl Acids Res 15:529–542

Dried Samples: Hard Tissues
14 Y-Chromosomal DNA from Ancient Bones

SUSANNE HUMMEL and BERND HERRMANN

1. Introduction

In prehistoric anthropology the sex ratio and sex-related mortality of a population are part of the basic biological data, alongside age structure, morbidity, and nutritional status. These data give access to sociocultural and socioeconomic characteristics of a population. While morphologic sex determination from the skeletal remains of adults is usually not a difficult procedure, anthropologists have lacked reliable methods of sex determination from the often physically damaged and fragile bones of infant individuals. Since medieval cemeteries may often contain up to 50% infant and juvenile burials, sex determination in subadult individuals is a necessity for paleodemography and related fields concerned with social history. In this chapter we will show how the analysis of ancient DNA (aDNA) can provide an important tool to these sciences by offering a molecular method of sex determination that is relatively independent of gross morphologic preservation.

The main sources of DNA in bone extractions are the following three types of bone cells: (i) osteocytes, which maintain bone metabolism and control the activity of (ii) osteoblasts, which are bone-forming cells, and (iii) osteoclasts, which are bone-eroding cells. Also present beside these major sources of DNA, which are easily detectable by histologic staining techniques even in ancient skeletal material, are blood cells (cf. Chapter 2 and Cattaneo et al. 1990) and epithelial cells within the Haversian systems, and probably residues of cartilage cells, at least in well-preserved specimens.

Since bone is a material lacking in liquids and enzymes, cells within bone can be expected to suffer less from autolytic processes; moreover, these cells are probably better protected against diagenetic influences than cells of any other tissue.

2. DNA Extraction

2.1 Sample Collection and Pretreatment

Sample collection should best be done immediately, while excavation is still in progress. This will save sample material from changes in environment which might further damage nucleic acid residues, e.g., when excavated skeletal material is washed and/or stored in a dry and warm climate. Instant sampling on site will also minimize the risk of contamination with modern human DNA.

Since opportunities for immediate sampling are often restricted or non-existent, one may need to work with skeletal material from museum collections instead, which is likely to have been contaminated with modern genomic DNA during excavation or previous investigations. Therefore, bone samples for molecular studies should be sections or slabs of compact bone with all outer surfaces either cut away or extensively UV irradiated. Due to the morphology of bone, compact bone samples combine maximum probability for good DNA preservation with minimal bacterial or fungal contamination.

Whether regular or automated extraction is perfomed, an amount of 0.5 g of bone will be sufficient for up to four extraction procedures and about 200 amplifications, respectively. The sample is broken into small pieces with a metal mortar before being ground to a fine powder in a mixer mill. About 0.5 g of this bone powder is transferred into a sterile tube and mixed with 0.5 M EDTA, pH 8.0, in order to decalcify the sample. Surplus bone powder may be stored at $-20°C$. With fine-powdered bone material a 48–72 h agitated incubation with EDTA will suffice for DNA extraction. Changing the EDTA solution to accelerate decalcification of the sample is not recommended, since due to diagenetic and mechanic cell damage free nucleic acids are found in solution very early on. Quantitative cell lysis, proteinase K digestion, phenol/chloroform extraction, and DNA concentration and purification are performed in the Genepure-341a nucleic acids extractor of Applied Biosystem (Darmstadt-Weiterstadt) (7 ml version). Details on manual extraction of bone are given in Hummel and Herrmann (1991).

2.2 Automated DNA Extraction

Before some of the cell suspension is removed remaining larger particles of solid matter should have been given some minutes to settle; alternatively, the suspension has to be briefly centrifuged. The processing of the ancient DNA (aDNA) in a nucleic acids extractor is then done as follows:

Cell suspension (500 µl per sample) is added to well-silanized glass tubes containing 500 µl lysis buffer each. After brief mixing, the tubes are incubated and slowly agitated for 1 hr at 60°C with 50 µl proteinase K (20 mg/ml) each. An extraction is performed once with an approximately equal volume of phenol/chloroform (1 ml) by thorough agitation for 6 min at room temperature.

Phases are allowed to separate while the samples rest for 8 min at 60°C. One chloroform extraction (1 ml) is then performed by agitation for 6 min followed by phase separation at 60°C for another 6 min.

2.3 Concentration and Purification

The concentration and purification of the extracted DNA combines three strategies: alcoholic precipitation of DNA, binding of DNA to silicate glass beads, and retention of DNA by membrane filtration. After extraction (cf. above) each sample consists of an approximate volume of 6 ml. To facilitate precipitation and binding of the DNA to glass milk, 2 M sodium acetate (500 µl) is added to the sample. An equal volume of isopropanol is added and mixed with the sample for 75 sec before 5 µl glass milk (Dianova, Hamburg) is added. Binding of the DNA to glass milk is allowed to take place for 10 min under moderate agitation at room temperature. Finally the solution is filtered through a membrane consisting of either teflon or cellulose, with pore size ranging from 3 to 5 µm. Only the DNA–glass milk complex is held back in this step. It is washed once with absolute ethanol for 5 min; in this procedure salts and humic acid impurities are usually removed quantitatively. The DNA–glass milk complex is washed off the membrane with 80% ethanol, which is removed afterwards by centrifugation. After drying of the DNA–glass milk pellet, the DNA is eluted with 50 µl of sterile water.

3. Control Samples for Sex Determination

In the sexing of individuals by means of the PCR technique, control samples are needed to serve two functions. First, one has to take into account the risk of false-positive results due to modern contamination by either male whole genomic DNA or product carryover (cf. Chapter 4); second, one needs to recognize false-negative results caused by inhibition of the reaction. There are three types of controls which will effectively cover these possibilities.

3.1 Positive Controls

A positive control is prepared from DNA of a male individual; its chemical composition should be as close as possible to the samples to monitor for false-negative results. Therefore, control DNA is best extracted from bone material of a definitively male individual. If dissection material is used, extraction and reaction parameters are also controlled for. If a series of samples from, for example, a (pre-)historic cemetery is investigated, an ancient male sample may furthermore control against PCR inhibitors such as microorganismal DNA and humic acids; thus an ancient male sample can work as an internal control if, for example, the amount of target DNA introduced to the amplification reaction must be varied.

3.2 Negative Controls

DNA extracts of bones of female individuals are negative controls and will reveal possible false-positive results. False positives may be the result of nonstringent reaction parameters or of contaminating DNA sequences introduced to the samples during laboratory work. If ancient female samples are available from investigation of the same excavation site, these samples will allow the researcher to check the effectiveness of decontamination sample pretreatment (see above).

3.3 No-Template and Blind Controls

These types of controls do not contain sample DNA and are the most effective controls against contaminations of the carryover type. They either are pure 0.5 M EDTA run through the complete extraction procedure (blind control) or are prepared as usual for PCR amplification but with sterile water instead of DNA (no-template control).

4. DNA Amplification

4.1 Human Y-chromosome-Specific Primers

There are two DNA sequences and respective pairs of primers known to be specific for the human Y-chromosome. One is a pair of primers (Y1/Y2) that spans a 170-bp fragment (Witt and Erickson 1989) of a 5.5-kb repeat sequence of the α-satellite region of the Y-chromosome (Wolfe et al. 1985). Each Y-chromosome carries about 100 copies of this sequence. The other is a pair of primers (JR25/JR26) that amplifies a 154-bp fragment (Kogan and Gitschier 1990) of a 3.4-kb repeat sequence on the long arm, which is copied 800 to 5,000 times on each Y-chromosome (Nakahori et al. 1986). Both pairs of primers are familiar from forensic applications or prenatal diagnostics. The sequences of the oligonucleotides are:

```
Y1:    5′ ATG ATA GAA CGC AAA TAT G    3′
Y2:    5′ AGT AGA ATG CAA AGG GCT CC   3′

JR25: 5′ TCC ACT TTA TTC CAG GCC TGT CC 3′
JR26: 5′ TTG AAT GGA ATG GGA ACG AAT GG 3′
```

4.2 Amplification Parameters

Reaction mixtures should be set up as master mixes containing all components except *Taq* polymerase, in order to avoid concentration differences in reaction components for a given series of samples. Exclusion of one essential PCR component from the prepared mixture, to be only introduced to the reaction once the denaturing temperature is reached within the first amplification cycle, prevents unspecific elongations due to possible primer hybridization to nontarget DNA sequences at lower (room) temperatures (Ruano et al. 1991).

Standard concentrations of reaction components for the amplification of Y-specific sequences with Y1/Y2 and JR25/JR26 are 10 mM Tris-HCl pH 8.3, 50 mM KCl, 1.5 mM MgCl$_2$, 200 μM dNTPs, and 5–10 pM of each primer. For a 50-μl reaction volume 0.5–1 μl DNA (1–2% of the DNA extracted as described above) should be used as template. The reaction mix now containing DNA is overlaid with sterile mineral oil and exposed to a first, extended period of denaturing temperature (94°C, 5 min). During this period, 2 U *Taq* polymerase per sample is added.

Typical amplification parameters are: 30 cycles at 94°C (1.5 min), 58°C (20 sec), 72°C (1 min). If the primers for the 170-bp fragment are used, it is recommended to decrease the annealing temperature from 58°C to 55°C because of the lower hybridization optimum of these shorter primers; it will also be regularly necessary to increase the number of amplification cycles from 30 to 40, since the target sequence is expected to be present less often in comparable volumes of DNA extract.

If the amplification reaction is performed with DNA extracts of lower purity, as is often the case for manual extractions, it might be necessary to run a biphasic PCR. For details of amplification parameters in biphasic amplification procedures see Hummel et al. (1992).

5. DNA Analysis

5.1 Electrophoresis

The presence or absence of Y-chromosome-specific amplification products on ethidium bromide stained agarose gels leads to sex determination. However, the absence of product will not always prove beyond doubt that the individual under investigation is female. It could also be due to the DNA extract not being amplifiable for some reason (e.g., DNA fragments that are too short, or presence of inhibitors). Since there is no positive evidence for female sex, other, non-sex-related primers which amplify fragments of comparable length should be applied to such samples to demonstrate the amplifiability of the sample extract.

5.2 Sequence Analysis

To prove the specificity of amplification products we carried out sequence analyses of reactions of both male and female samples that had been processed identically for cloning. Only the PCR products of the males revealed inserts by blunt-end ligation into a plasmid vector and transformation into a competent *E. coli* strain. After the preparation of single-stranded DNA, automated sequencing with internal fluorescent labeling was performed (Hummel et al. 1992). The expected 154-bp sequence of the PCR products could be confirmed for modern and ancient samples (cf. cover illustration of this volume).

6. Conclusion

In our experience the automated extraction of DNA from ancient samples yields higher and more fully reproducible quantities of DNA from a given sample volume than does manual extraction. Moreover, the quality of the DNA extracts is improved by automated extraction, i.e., the DNA shows a degree of purity which is almost comparable to that of modern samples. Under these improved conditions it is possible to reduce the total number of PCR cycles drastically and to return to monophasic amplifications. The main advantage of this lies in the reduced risk of contamination by carryover and the reduced cost of PCR as less *Taq* polymerase is required.

As with any processing of aDNA, the choice of proper control samples is one of the basic prerequisites for successful sex determination. In doubtful cases control amplifications with non–sex-dependent primers and/or sequencing of PCR products will help to interpret results.

Acknowledgment. This work has been supported by a grant of the German Bundesminister für Forschung und Technologie.

References

Cattaneo C, Gelsthorpe K, Phillips P, Sokol RJ (1990) Blood in ancient human bone. Nature 347:339

Hummel S, Herrmann B (1991) Y-chromosome-specific DNA amplified in ancient human bone. Naturwissenschaften 78:266–267

Hummel S, Nordsiek G, Herrmann B (1992) Improved efficiency in amplification of ancient DNA and its sequence analysis. Naturwissenschaften 79:359–360

Kogan SC, Gitschier J (1990) Genetic prediction of Hemophilia A. In: Innis MA, Gelfand DH, Sninsky JJ, White TJ (eds) *PCR Protocols: A Guide to Methods and Applications.* Academic Press, San Diego: Sec. 288–299

Nakahori Y, Mitani K, Yamada M, Nakagome Y (1986) A human Y-chromosome specific repeated DNA family (DYZ1) consists of a tandem array of pentanucleotides. Nucl Acids Res 14:7569–7580

Ruano G, Brash DE, Kidd KK (1991) PCR: the first few cycles. Amplifications 7:1–4

Witt M, Erickson RP (1989) A rapid method for detection of Y-chromosomal DNA from dried blood specimens by the polymerase chain reaction. Hum Genet 82:271–274

Wolfe J, Darling SM, Erickson RP, Craig IW, Buckle VJ, Rigby PWJ, Willard HF, Goodfellow PN (1985) Isolation and characterization of an alphoid centromeric repeat family from the human Y chromosome. J Mol Biol 182:477–485

Dried Samples: Hard Tissues
15 Genomic DNA from Museum Bird Feathers

HANS ELLEGREN

1. Introduction

Ancient avian DNAs have been analyzed from preparations of skins of museum specimens (Houde and Broun 1988; Arctander and Fjeldså 1991; Smith et al. 1991), from soft tissue remains (Houde and Braun 1988; Cooper et al., 1992), and from bones (Cooper et al., 1992). Principally the questions that can be addressed by a molecular approach to ancient DNA (aDNA) do not differ from birds to, for instance, mammals. The large number of birds collected and stored at museums worldwide particularly facilitates studies that aim to investigate the genetic variability of past avian populations or their population dynamic and distribution. However, for large-scale sampling of old museum birds it would be advantageous to use a noninvasive sampling method, i.e., one that does not involve destruction of the skin. Successful PCR amplification from single hairs (Higuchi et al. 1988) suggested that it should be possible to amplify DNA from other keratinic tissues such as feathers. Indeed, this has recently been demonstrated by feather amplifications of mitochondrial DNA (mtDNA) sequences (Smith et al. 1991; Taberlet and Bouvet 1991) as well as of genomic sequences (Ellegren, 1992) from contemporary birds. Moreover, Ellegren (1991) has shown that genomic sequences can be amplified and analyzed from a single, more than 100-year-old museum feather.

In this chapter I will discuss the amplification of hypervariable genomic sequences (microsatellites) in DNA prepared from single feathers of museum birds. Analyzing genetic variability at such highly polymorphic nuclear loci (DNA typing) is a rapid approach to the study of genetic diversity in populations.

2. Hypervariable Microsatellite Markers

Microsatellites, or simple repeat sequences, are tandem reiterations of mono-, di-, tri-, or tetrameric nucleotide motifs. Their abundance in eukaryotic genomes was demonstrated in hybridization experiments about ten years ago by Hamada and coworkers (e.g., Hamada et al. 1982). The most frequent motif, $(GT)_n$, probably occurs as a repetitive element interspersed every 30 kb in the human genome (Stallings et al. 1991). Cloning of individual microsatellite loci has shown that they typically consist of 10–30 tandem copies of the repeat motif. Sequences flanking microsatellites are unique, and by designing primers to each side of the repeat, locus-specific 100–300-bp products can be obtained. Microsatellites display a significant degree of polymorphism which is mainly due to allelic variation in the number of tandem repeats (Litt and Luty 1989; Tautz 1989; Weber and May 1989). Heterozygosities of 80–90% are not uncommon. Alleles are inherited in a stable Mendelian fashion and this, together with their high frequency and hypervariability, makes microsatellites highly suitable as markers for linkage analysis or for forensics. Recently, the first avian microsatellites have been characterized (Ellegren, 1992).

The variable nature of microsatellites also suggests that they would be useful for studies of genetic variability in wildlife populations, analogous to the use of minisatellites/DNA fingerprinting for the same purpose (e.g., Gilbert et al. 1990). An interesting feature in the context of ancient studies is that microsatellites constitute much shorter sequences than minisatellites and are thus more likely to have survived intact in degraded aDNA (Ellegren, 1991; Hagelberg et al. 1991).

3. Techniques

3.1 Use of Chelating Resins

Some electron-rich compounds such as aminosalicylic acid, 8-hydroxyquinoline, carboxymethylated amino acids, and iminoacetic acid are characterized by their ability to form chelating complexes with polyvalent transition metals. These complexes are due to ion–dipole interactions in which the metal coordinates molecules containing O, N, or S. Some amino acids show affinity for heavy metals, a fact exploited in metal chelate chromatography (Porath et al. 1975). In this variant of affinity chromatography, a chelating agent is immobilized on an insoluble support (resin). The most widely used agent is iminodiacetic acid, which forms different types of multicoordinated metal complexes depending on the spatial configuration in the support.

The capacity of chelating resins to form complexes with metals can also be of advantage in DNA preparation. Since an ultrapure DNA sample is not a prerequisite for enzymatic amplification, the main point of the pre-PCR treatment is to make DNA molecules accessible to the *Taq* polymerase. Cell disruption and protein denaturation by boiling liberates DNA molecules and inactivates, e.g., proteases but has the disadvantage that DNA may be degraded, a process likely to be catalyzed by metal ions. It has been suggested that by adding chelating ions before boiling, DNA degradation can be prevented (Singer-Sam et. al. 1989).

One type of chelating resin used for DNA preparation is Chelex 100 (BioRad). This matrix is composed of styrene–divinylbenzene copolymers (150–300 μm beads) containing iminodiacetate ions. As a rule, the sample is boiled in 5% Chelex solution, and fractions of the supernatant are then subjected to PCR.

3.2 DNA Preparation

DNA can be prepared and amplified from a variety of mature feather types including primaries, secondaries, rectrices, and body contour feathers. There is no need to use blood quills.

A single feather is plucked from a specimen by grabbing the lower part of the main shaft (rachis), making sure that the basal section (calamus: where pulp cells are to be found; Stettenheim 1972) is collected. Since old feathers may be fragile it is advisable to use forceps. Approximately 5–10 mm of calamus is cut into millimeter-sized pieces with a surgical blade. Separate tools must be used for each specimen.

Feather pieces are transferred to a 5% Chelex-100 solution containing 0.1 μl proteinase K. This can be done in 200 μl; however, if one expects no more than minute yields and plans to use the entire preparation for a single PCR reaction, the volume can be decreased. It must be emphasized that when pipetting Chelex solutions, the beads must be evenly distributed in the solution. This is achieved most easily with a magnetic stirrer and a stir bar. Pipetting also requires a relatively large bore of the tip; if a typical 200-μl tip is used, the very end may have to be cut off.

The mixture is incubated at 56°C for at least 2 hr, after which an additional portion of proteinase K may be added and incubation continued. Subsequently, the sample is vortexed for 10 sec, boiled for 8 min, and then vortexed again. The beads are pelleted (2 min in a bench centrifuge at 13,000 rpm) and the DNA-containing supernatant subjected to PCR.

The remaining solution can be frozen; before it is used again, it should be vortexed and centrifuged. Alternatively, the entire supernatant can be removed and frozen for later use.

3.3 PCR Amplification

For a 20-μl reaction, 10 μl Chelex supernatant is added to 10 μl of a PCR mix containing 1.5 mM $MgCl_2$, 50 mM KCl, 10 mM Tris-HCl pH 8.0, 2 μg/ml bovine serum albumin (BSA), 200 μM dNTP, 0.5 U *Taq* polymerase, and 2–10 pmole of each microsatellite flanking primer. The inclusion of BSA in the PCR reaction significantly improves the results, a situation also found in preparations from other ancient tissues (e.g., Pääbo 1990).

PCR is carried out with cycles of 94°C for 1 min, 55°C for 30 sec, and 72°C for 1 min. Normally it is necessary to perform 40-cycle reactions. In the initial cycle(s) a prolonged denaturation step (e.g. 95°C for 2 min) is recommended, whereas the last cycle should be followed by an extra 5–10-min extension. To compensate for inactivation of the *Taq* polymerase during repeated heating, either more enzyme can be added after 30 cycles, or the entension step can be increased by 1–2 sec on every cycle.

In DNA preparations from ancient tissues or bones, substances inhibiting the activity of the *Taq* polymerase have sometimes been observed (Pääbo 1990). In a few cases this has also been noted for feather amplifications. Since there are various ways to prepare skins for museum storage (e.g., treatment with arsenic or organic compounds), the ability of DNA samples from museum feathers to undergo PCR may be related to the skin treatment history of the specimen (cf. Cooper, Chapter 10).

The risk of contamination can never be ignored in any kind of PCR analysis. A negative control and measures to avoid sample carryover are essential (Kwok 1990). In the case of museum bird feathers, two feathers plucked from different parts of the bird should be independently analyzed. Regarding heterologous contamination, it should be noted that primers flanking microsatellite loci may amplify the corresponding sequence in closely related species (Moore et al. 1991). However, it is highly unlikely that avian primers would amplify human sequences, due to the large evolutionary distance between the two lineages (cf. Ellegren 1991).

3.4 Analysis of Microsatellite Polymorphism

Microsatellite polymorphism, due mainly to length variations in the repeat region, permits alleles to be distinguished by electrophoretic separation. Since two alleles may differ with regard to just a single repeat unit, i.e., show a size divergence of only one or a few nucleotides, it is usually necessary to run long denaturing polyacrylamide (sequencing) gels and visualize the alleles by autoradiography. However, if alleles show significant divergence in size, they may be distinguishable by agarose gel electrophoresis and subsequent ethidium bromide staining. An example of microsatellite typing of museum birds is given in Figure 1.

Radioactive assays are carried out by incorporating ^{32}P-dCTP during PCR strand synthesis or, preferably, by end-labeling one of the primers with ^{32}P-dATP. A standard protocol for end-labeling can be found in Sambrook et al. (1989). Signals of suitable strength are obtained when using 0.5 µCi ^{32}P-dATP per picomole primer in the labeling reaction.

PCR products are mixed with formamide and dye, heated to 80°C, and loaded on a 6% denaturing polyacrylamide gel. To facilitate the identification of alleles, a sequence ladder can be run adjacent to the amplifications.

4. Perspective

The only previously reported PCR amplification of genomic DNA sequences from single feathers of museum birds allowed for the analysis of microsatellite polymorphism in two passerine species, the pied flycatcher *Ficedula hypoleuca*

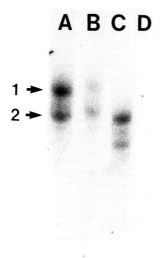

FIGURE 1. PCR amplification of the polymorphic microsatellite locus PTC3 in three museum specimens of flycatchers. Lane *A* is a pied flycatcher (*Ficedula hypoleuca*) collected in Sweden in 1908 (Swedish Museum of Natural History, no. 10) that is homozygous for allele 1, lane *B* is an individual of the same species collected in Finland in 1910 (no. 2) that is either heterozygous 1/2 or homozygous for 1, lane *C* is a Collared flycatcher (*Ficedula albicollis*) collected in Sweden in 1905 (no. 2) that is homozygous for allele 2, and lane D is a negative control using only the common PCR master mix. DNA was prepared from a single body feather of each individual. Primer sequences were: PTC3A: 5'-GTGTTCTTAAAACATGCCTGGAGG; PTC3B: 5'-GCACAGGTAAATATTTGCTGGGCC (Ellegren, 1992).

and the swallow *Hirundo rustica* (Ellegren 1991). The oldest bird analyzed was collected in Sweden in 1860. Microsatellite markers used as tools for the assessment of genetic variability will now make possible comparative studies of past and contemporary populations. Such studies are of considerable importance in the context of conservation programs for endangered populations. If it can be shown that a population has lost genetic variability during its decline, this demonstrates the need for introduction of new genetic material. Although any finding of a low degree of genetic variability in a small population is a strong indication of loss of diversity, this cannot be conclusively associated with a population bottleneck unless reference is made to the genetic diversity when the bottleneck was developed.

This chapter has focused on amplifications of hypervariable genomic loci from ancient feathers. There are good reasons to believe that other sequences (e.g., mtDNA) can also be amplified from museum feathers (Smith et al. 1991; Taberlet and Bouvet 1991), allowing several types of studies to be initiated. The fact that feather analysis implies an almost nondestructive sampling method speaks in favor of this approach for genetic studies of museum birds.

Acknowledgments. This study was given financial support by the Uddenberg-Nordingska and Nilsson-Ehle Foundations. Ulf Gyllensten kindly gave advice on DNA preparation.

References

Arctander P, Fjeldså J (1991) DNA studies for avian systematics: techniques, virtues and perspectives of DNA collections. In: Edelstam C, Mina M (eds) *Museum Research in Vertebrate Zoology.* Stockholm: Swedish Museum of Natural History

Cooper A, Mourer-Chauviré C, Chambers GK, von Haeseler A, Wilson AC, Pääbo S (1992) Independent origins of New Zealand moas and kiwis. Proc Natl Acad Sci USA 89:8741–8744

Ellegren H (1991) DNA typing of museum birds. Nature 354:113

Ellegren H (1992) Polymerase chain reaction (PCR) analysis of avian microsatellites: A new approach to studies of genetic relationships in brids. Auk 109:886–895

Gilbert DA, Lehman N, O'Brien SJ, Wayne, RK (1990) Genetic fingerprinting reflects population differentiation in the California channel island fox. Nature 344:764–767

Hagelberg E, Gray IC, Jeffreys AJ (1991) Identification of the skeletal remains of a murder victim by DNA analysis. Nature 352:427–429

Hamada H, Petrino MG, Kakunaga T (1982) A novel repeated element with Z-DNA forming potential is widely found in evolutionarily diverse eukaryotic genomes. Proc Natl Acad Sci USA 79:6465–6469

Higuchi R, von Beroldingen CH, Sensabaugh GF, Erlich HA (1988) DNA typing of single hairs. Nature 332:543–546

Houde P, Braun MJ (1988) Museum collections as a source of DNA for studies of avian phylogeny. Auk 105:773–776

Kwok S (1990) Procedures to minimize PCR-product carry-over. In: Innis MA, Gelfland DH, Sninsky JJ, White TJ (eds) *PCR Protocols: A Guide to Methods and Applications.* San Diego: Academic Press, pp. 142–145

Litt M, Luty JA (1989) A hypervariable microsatellite revealed by in vitro amplification of a dinucleotide repeat within the cardiac muscle actin gene. Am J Hum Genet 44:397–401

Moore SS, Sargeant LL, King TJ, Mattick JS, Georges M, Hetzel DJS (1991) The conservation of dinucleotide microsatellites among mammalian genomes allows the use of heterologous PCR primer pairs in closely related species. Genomics 10:654–660

Pääbo S (1990) Amplifying ancient DNA. In: Innis MA, Gelfland DH, Sninsky JJ, White TJ (eds) *PCR Protocols: A Guide to Methods and Applications.* San Diego: Academic Press, pp. 159–166

Porath J, Carlsson J, Olsson I, Belfrage G (1975) Metal chelate affinity chromatography: a new approach to protein fractionation. Nature 258:598–599

Sambrook J, Fritsch EF, Maniatis T (1989) *Molecular Coloning: A Laboratory Manual,* 2nd ed. Cold Spring Harbor, N.Y.: CSHL Press

Singer-Sam J, Tanguay RL, Riggs AD (1989) Use of Chelex to improve the PCR signal from a small number of cells. Amplifications 3:11

Smith EFG, Arctander P, Fjeldså J, Amir OG (1991) A new species of shrike (*Laniidae: Laniarius*) from Somalia, verified by DNA sequence data from the only known individual. Ibis 133:227–235

Stallings RL, Ford AF, Nelson D, Torney DC, Hildebrand CE, Moyzis RK (1991) Evolution and distribution of $(GT)_n$ repetitive sequences in mammalian genomes. Genomics 10:807–815

Stettenheim P (1972) The integument of birds. In: Farner DS, King JR (eds) *Avian Biology*, vol. 2. London: Academic Press, pp. 2–63

Taberlet P, Bouvet J (1991) A single plucked feather as a source of DNA for bird genetic studies. Auk 108:959–960

Tautz D (1989) Hypervariability of simple sequences as a general source for polymorphic DNA markers. Nucl Acid Res 17:6463–6471

Weber JL, May PE (1989) Abundant class of human DNA polymorphisms which can be typed using the polymerase chain reaction. Am J Hum Genet 44:388–396

Dried Samples: Hard Tissues
16 DNA and RNA from Ancient Plant Seeds

FRANCO ROLLO, FRANCO MARIA VENANZI, and AUGUSTO AMICI

1. Introduction

Plant seeds are peculiar structures designed to protect the genetic material of their embryo until long after the death of the mother plant. For this reason, the cells of a plant seed are adapted to survive the most stringent environmental conditions. After prolonged storage, however, a seed loses its capacity to germinate (viability). If the seed is kept in a dry environment it undergoes a sort of spontaneous mummification with no appreciable change in its external morphology. With the passing of centuries it will acquire a dark reddish-brown color, yet show an amazing degree of structure conservation (for a review, see Toole 1986).

Archaeological findings of ancient seeds tell us much about ancient civilizations and, in particular, about the stages through which the domestication of crop plants took place (Zohary and Hopf 1988). Nevertheless, the traditional archaeobotanical approach, essentially based on the study of ancient seed morphology, has proved unable to give satisfactory answers to the many questions posed by the debate on plant domestication; hence the reiterated attempts in the past to utilize seed storage protein profiles to establish relationships between ancient and present-day taxa. None of these studies, however, has given convincing evidence that such an approach is feasible (Shewry et al. 1982). More recently, the attention of researchers has focused on the possibility of isolating DNA from ancient seeds and other plant remains. An increasing number of experimental results indicate that endogenous nucleic acids (RNA and DNA) can indeed be found in mummified seeds. This DNA is usually fragmented into small pieces and only present in low amounts compared to modern seeds. But despite these problems, proper molecular research strategies which take into account both the state of preservation of the DNA and the mode of evolution of the different genomes of the plant cell allow for significant genetic information to be retrieved.

2. Historical Background

Early attempts to isolate nucleic acids from ancient plant seeds were carried out in the seventies by Daphne Osborne and coworkers. Though the ultimate goal of their research was to explain the loss of viability in seeds, rather than to retrieve ancient genetic information, they addressed an issue fundamental to all modern molecular genetic investigations on ancient seeds, i.e., the fate of the DNA during the aging process. According to these authors, DNA fragmentation takes place very soon after seed death: 100-year-old seeds contain highly degraded ribosomal RNA (rRNA) and DNA already, and only very low molecular weight nucleic acids can be found in neolithic grains from Egyptian tombs (Cheah and Osborne 1978; Osborne, pers. comm.). More recently, Rollo (1985) analyzed old cress (*Lepidium sativum* L.) seeds from the Thebes necropolis. These were found to contain low molecular weight, ribonuclease-sensitive nucleic acids that hybridized to cloned plant genes for rRNA but not to bacterial DNA. In the same year Rogers and Bendich (1985) reported the extraction of polymerized DNA from mummified seeds and seed embryos ranging in age from 500 to more than 44,000 years. However, they provided no evidence for an endogenous origin of this DNA. The first clear evidence for the presence of endogenous DNA in an ancient plant seed specimen was given by Rollo et al. (1988) by use of the polymerase chain reaction (PCR).

3. Nucleic Acid Extraction and Characterization

3.1 Extraction Procedures

The isolation of nucleic acids from mummified plant seeds may require several steps, such as external washing of the seed remains to remove contaminating soil particles, fungal hyphae etc., milling of the seeds, and extraction of the resulting powder using appropriate solvents. The extracted nucleic acids can be purified further by standard laboratory techniques such as gel electrophoresis fractionation. The first step, i.e., external washing to remove absorbed contaminants, may not always be necessary. Also, as an alternative to milling, seeds can be left to soak and then be homogenized.

Nucleic acids can be extracted from flours or homogenates using phenol-based mixtures (Rollo 1985; Rollo et al. 1987, 1991) followed by phenol/chloroform mixtures and then by chloroform. Eventually, the nucleic acids are ethanol precipitated and resuspended in a suitable medium. This procedure produces nucleic acid pellets that are invariably dark brown stained, due to coprecipitating impurities. Most of the contaminants can be eliminated by agarose gel electrophoresis fractionation of the nucleic acids. This technique takes advantage of two facts: (i) under the standard electrophoretic conditions employed for the fractionation of nucleic acids, i.e., use of 1.5–2.5% agarose, the brown contaminants migrate slower than depolymerized ($< 200\,bp$) DNA and RNA; (ii) when exposed to UV light, the brown contaminants produce a light blue fluorescence which can easily be distinguished from the reddish fluorescence of the ethidium bromide stained nucleic acids. Those portions of the gels corresponding to the reddish fluorescence can then be excised and be either melted and submitted to a phenol extraction procedure to recover the nucleic acids, or directly added to the PCR reagents (see Methods below). A possible alternative is the CTAB (cetyltrimethylammonium bromide) nucleic acid extraction method of Rogers and Bendich (1985).

3.2 Biochemical and Immunological Characterization

A fundamental question is that of the precise nature of the nucleic acid fractions isolated from ancient seed samples. A first approach can consist in the analysis of their base content. Nucleic acids are extracted from the seeds, gel purified, submitted to hydrolysis using formic acid, then to high performance liquid chromatography (HPLC). The elution profile of the bases thus derived from a sample of 3,400-year-old cress (Rollo et al. 1991) shows sharp C, U, G, and A peaks and only a minor T component, compared to the profile of a standard mix of bases. Several comments can be made on the origin of the uracil peaks in the ancient nucleic acid HPLC profile. First, uracil peaks cannot originate from unpolymerized uracil in ancient seeds, as HPLC profiles are obtained from gel-purified polymerized material. Second, although in principle uracil can be produced by deamination of cytosine, experiments performed by Pääbo (1989) have shown that ancient nucleic acids are only to a limited extent sensitive to uracil DNA glycohydrolase (an enzyme that removes uracil from DNA), suggesting that deamination of cytosine plays a minor role in the postmortem modification of DNA. Finally, cytosine is rather stable in the medium employed to hydrolyze nucleic acids (concentrated formic acid), which rules out the possibility of uracil arising from degradation of cytosine during hydrolysis (Chargaff 1950).

All this indicates that RNA makes up a major proportion of the nucleic acid fractions isolated from ancient plant seeds. Additional evidence in this respect comes from the immunological characterization of the nucleic acid fractions. Antibodies against double-stranded (ds) DNA are often detected in the sera of patients with systemic lupus erythematosus (SLE) disease. Detailed investigations on antigenic determinants for SLE anti-DNA antibodies have shown a lack of direct involvement of the bases in the antigen–antibody reaction, and demonstrated instead the capacity of these antibodies to distinguish between 2-deoxyribose-phosphate and ribose-phosphate backbones (Stollar 1986). Antibodies to dsDNA may also arise spontaneously in normal individuals and animals, and in recent years hybridomas have been generated to produce a variety of "natural" anti-DNA antibodies. When nucleic acid fractions obtained from cress seeds are made radioactive and treated with anti-DNA IgG from the sera of SLE patients, or with monoclonal anti-DNA IgM from mouse, one can observe in both cases that only a minor proportion

($\leqslant 0.2$–0.3%) of the ancient nucleic acids is precipitated. This result is in sharp contrast with what is found when modern low molecular weight DNA is subjected to the same treatment, either alone or in combination with nucleic acids from ancient seeds (Rollo et al. 1991).

In conclusion, ancient plant seeds contain degraded rRNA and DNA. The RNA component, stemming mainly from fragmentation of the large 25S and 18S cytoplasmic rRNAs (Rollo et al. 1991), is by far the more abundant of the two. This phenomenon simply reflects the situation in the living seed, where rRNA constitutes $> 90\%$ of the nucleic acid content. In nucleic acid preparations obtained from ancient seeds, DNA can be detected and characterized by the use of PCR (Rollo et al. 1988).

4. Microbial Contamination

Contamination of ancient remains, whether of human, animal, or plant origin, by modern microorganisms should always be considered a fairly likely occurrence. In the case of ancient plant seeds dispersed in soil it is conceivable that the microbiol flora of the soil eventually penetrated into the seed and utilized the seed cellular structures as support and substrate for growth.

It is well known that microbial nucleic acids can be extracted with good yield from soil samples (Ogram et al. 1988); thus, the finding per se of DNA and RNA associated with the remains of ancient seeds can never be taken as evidence that some of the original genetic material of the seeds has been preserved. Indeed, several studies have provided clear evidence for microbial contamination of ancient seed remains. R. Colwell found bacteria in scanning electron micrographs of figs and olives from archaeological sites at Pompeii and Herculaneum. Fungal spores were found by the same scholar in carbonized hay at Torre Annunziata (Meyer 1980).

Seed samples of *Ceratophyllum* sp., *Cirsium* sp., *Ficus carica* L., *Ranunculus* sp., *Rubus* sp., *Vitis vinifera* L., and *Zannichellia* sp. from the archaeological site of the "sacred spring" of Pizzica Pantanello near Metaponto, a Greek colonial city in southern Italy, were analyzed by Rollo et al. (1987) and, despite a remarkable state of preservation, were found to contain bacterial and fungal contaminants. According to archaeologists, the remarkable state of preservation of the seeds was attributable to the "anaerobic conditions created by the ground water, flowing as it had in ancient times from the mouth of the spring and into the reservoir" (Toole, 1986). *Ranunculus* sp. seeds found in the mid fourth century level indicated a humid but not thoroughly submerged environment, whereas the presence of *Ceratophyllum* sp. and *Zannichellia* sp. indicated a swampy condition creating a very compact stratum of organic material in the level of the immediately succeeding period. Radiocarbon-dated (2200 ± 60 to 2320 ± 70 years B.P.) seed samples were submitted to nucleic acid extraction, and the nucleic acids were characterized by agarose gel electrophoresis. All samples were shown to contain polymerized DNA, up to 10–20 kb, a situation reminiscent of that described by Rogers and Bendich (1985). However, when these nucleic acids were tested by molecular hybridization using Southern blots, it was observed that cloned rDNA from plants did not bind to the DNA isolated from the ancient seeds under stringent hybridization conditions. As an additional control, the DNA isolated from one of the *Rubus* sp. samples was further purified, partially cut with the restriction enzyme *Sau*3A1 and cloned into plasmid pBR322. Six recombinant clones carrying inserts ranging from 300 to 600 bp were made radioactive and used to probe DNA isolated from modern *Rubus* sp. seeds. None of the probes appeared to be capable of hybridizing significantly with the modern DNA. Similarly, agarose gel electrophoresis analysis of the nucleic acids isolated from *Rubus* sp. and *Vitis vinifera* seeds from the Lombard archaeological site of Castelseprio (northern Italy, 9th century A.D.) revealed highly polymerized DNA and apparently intact rRNA bands. However, when compared with modern standards, the rRNA bands were shown to migrate as prokaryotic 23S and 16S rRNAs and not as 18S and 25S plant rRNAs. In addition, a scanning electron microscope survey of the ancient *Rubus* sp. seeds unequivocally revealed the presence of microbial contaminants.

Poorly preserved soil-contaminated pre-Columbian maize kernels from the archaeological site of the Sipan pyramid on the coast of Peru (Moche culture, approx. fifth century A.D.) were shown to contain polymerized DNA (up to about 1,000 bp). However, when made radioactive and used as a probe in a reverse Southern blot experiment, the DNA isolated from the ancient maize did not significantly hybridize with cloned plant rDNA. The same DNA was found to be refractory to PCR amplification using oligonucleotides designed to bind to short (90–130 bp) maize-specific nuclear and mitochondrial sequences (Rollo et al. 1991).

In conclusion, the isolation of high molecular weight nucleic acids from an ancient seed sample should alert the experimenter to the possibility of microbial contamination of the sample. The presence of depolymerized nucleic acids, however, may represent a first encouraging indication that some of the original genetic material of the seed survived (Table 1).

5. Choosing a Plant DNA Sequence for Diachronic Investigation

5.1 Mitochondrial DNA

Though it may seem trivial, we believe it is worth stressing that prior to undertaking any attempt to retrieve genetic information from ancient remains, one has to choose very carefully the DNA sequence to be searched for. This choice should take into account both the mode of evolution of the organism

TABLE 1. Endogenous versus contaminating nucleic acids in plant seeds from archaeological excavations

Species	Source	Age (year)	Endogenous nucleic acids	Bacterial contamination
Vitis vinifera	Shahr-i-Sokhta (Iran)	4,000–5,000	−	−
Lepidium sativum	Thebes (Egypt)	3,400	+	−
Rubus sp.	Castelseprio (Italy)	1,200	−	+
Vitis vinifera	"	"	−	+
Ceratophyllum	Metaponto (Italy)	2,200–2,300	−	+
Cirsium sp.	"	"	−	+
Ficus sp.	"	"	−	+
Ranunculus sp.	"	"	−	+
Rubus sp.	"	"	−	+
Vitis vinifera	"	"	−	+
Zannichellia sp.	"	"	−	+
Zostera marina	"	"	−	+
Zea mays	Huari site (Peru)	1,000	+	−
Zea mays	Moche site (Peru)	1,500	−	+

under investigation and the state of preservation of the DNA in the available sample. Whatever the ultimate aim of the diachronic investigation, in practice it will involve a search for genetic polymorphism at the molecular level.

A traditional target for molecular diachronic investigations in humans and animals is mitochondrial DNA. Mitochondrial DNA (mtDNA) is present in several hundred copies per cell and is transmitted to the progeny uniparentally, rendering relatively simple the interpretation of evolutionary or population data. In addition, mitochondrial tracts such as the D-loop region show considerable intraspecific divergence. Because of these properties, mtDNA has proved an invaluable system for analyzing ancient human and animal remains (Higuchi et al. 1987).

Unfortunately, these considerations apply only in part to plants. Although plant mtDNA is most often inherited uniparentally and present in high copy number per cell, it is known to undergo a peculiar evolutionary pattern: it evolves rapidly in structure, through numerous internal rearrangements, but slowly in sequence. Recent estimates suggest that the point mutation rate in plant mtDNA can be 100 times slower than that in animal mtDNA (Palmer and Herbon 1988). As a consequence, analysis of brief tracts of plant mtDNA, such as those which can be recovered from ancient seed remains, is not expected to show appreciable polymorphism when evolutionarily related organisms are compared. An example for this is the analysis of the extant members of the genus *Zea* performed by Rollo (1989).

5.2 Plastid DNA

The chloroplast (cp) genome of flowering plants consists of circular DNA molecules between 120 and 217 kb in size. With the exception of that of conifers and a group of legumes, cpDNA contains both large and small unique sequence regions which are separated by pairs of inverted repeats. The rate of sequence evolution of cpDNA is quite conservative as well. By some estimates the synonymous substitution rate is 0.1% per million years (Zurawski et al. 1984). Analysis of the distribution of restriction site mutations throughout the chloroplast genome indicates that the inverted repeats evolve more slowly than the unique sequence regions. Analysis of cpDNA was not found to be useful in discriminating between subspecies of *Zea mays* (Doebley et al. 1987). Moreover, while cpDNA is abundant in photosynthetic leaf tissues (10–20% of total cellular DNA), this is not the case for seed tissues, where plastid DNA accounts for only 1% of total cellular DNA.

5.3 Nuclear DNA

It is known that the overall rate of nuclear DNA evolution is about twice that of cpDNA, and specific nuclear gene sequences may evolve much faster still (Doebley et al. 1984). Though in principle, the relatively high rate of evolution of the nuclear plant genome will allow the researcher to discriminate among species even on the basis of very short tracts of DNA, limitations on the choice of sequence to be examined are posed by the fact that for ancient material, the detection of single-copy gene sequences requires many PCR cycles, which means a considerable risk of false results due to DNA carryover. Mid- or highly repetitive components of the nuclear genome offer an easier and more trustworthy target for DNA amplification than single-copy sequences do.

A particularly interesting parameter that can be utilized in molecular genetic investigations of ancient samples is the quantitative variation of repetitive components of the nuclear genome. In principle, unlike sequence variation polymorphism, copy number polymorphism should remain relatively unaffected by the fragmentation of the DNA which takes place during sample aging. Several examples of this type of polymorphism are known (Rivin et al. 1986).

Another interesting parameter that can be taken into account is the presence vs. absence of transposable element families, which represent a primary source of molecular evolution in higher plants. Transposable elements may be involved in both developmental and evolutionary processes, due to their capacity to initiate a whole range of chromosomal rearrangements, such as duplications, deletions, inversions, translocations, etc. In addition, plant transposable elements have been observed to generate sequence alterations at their site of integration. In *Zea mays*, many active transposable element systems have been described genetically and several have been characterized at the molecular level (Schwarz-Sommer et al. 1985).

6. Methods

Extraction of Nucleic Acids from Ancient Maize Kernels

For nucleic acid extraction, single kernels are put in a 10-ml test tube on a laboratory wheel and washed externally for 30 min at room temperature using 2 ml of mixture (phenol mixture) composed of 50 mM Na_2EDTA, 50 mM Tris-HCl pH 8.0, 1% (w/v) SDS, and 6% (v/v) water-saturated phenol. The tubes are removed from the wheel, the medium discarded and replaced with 1 ml fresh mixture, and seeds left to soak overnight at 4°C in the refrigerator. The soaked seeds are put in a sterilized mortar together with 1 ml fresh phenol mixture and homogenized with a pestle. The homogenate is extracted with 2 vol water-saturated phenol, vortexed for 2 min, spun for 1 min at 13,000 rpm in a benchtop centrifuge, and the aqueous supernatant is carefully collected and transferred to a new tube. Sodium chloride (1 M final) is added to the supernatant and the extraction procedure repeated using phenol:chloroform:isoamyl alcohol (24:24:1) and chloroform. To the last supernatant 0.6 vol isopropanol is added, and the resulting suspension kept in ice for 30 min. Nucleic acids are precipitated by centrifugation at 13,000 rpm for 10 min. The brown-stained nucleic acid pellet is desiccated, resuspended in 25 μl of sterile distilled water, then loaded on a $10 \times 7 \times 0.3$ cm 2.5% low-temperature agarose minigel containing 0.5 μg/ml ethidium bromide in $1 \times$ TBE ($1 \times$ TBE:90 mM boric acid, 2.5 mM Na_2EDTA, pH 8.0). The gel is run for 1 hr at 80 V, then put on a UV light transilluminator, and the whole fluorescent nucleic acid band excised and stored in a freezer.

Extraction of Nucleic Acids from Cress, *Rubus* sp., and Other Small Seeds

Seeds (5–20 seeds) are directly ground in a mortar with a pestle. The flours are resuspended in phenol mixture and extracted as described above.

Precise Evolution of the Degree of Polymerization of Nucleic Acids Isolated from Ancient Plant Seeds

Isopropanol-precipitated nucleic acids (10 ng) are terminally labeled using 3 units polynucleotide kinase (this enzyme is not inhibited by the brown contaminants) and 50 μCi $\gamma[^{32}P]$ATP for 30 min at 37°C in a final volume of 10 μl (Sambrook et al. 1989). Radioactive nucleic acids are precipitated twice and resuspended in a loading solution (95% formamide, 20 mM EDTA, 0.05% bromophenol blue, 0.05% xylene cyanol. Of this suspension 3 μl (approx. 2×10^6 cpm) is fractionated using a Sequi-Gen apparatus (Bio-Rad, Richmond, Cal.) on a $50 \times 21 \times 0.025$ cm 8% polyacrylamide gel with 5% crosslinking containing 7 M urea in $1 \times$ TBE. The gel is run for 2 hr at 40 W constant power. An accurate molecular weight standard is provided by the dideoxy chain termination sequencing reaction of a known plasmid DNA (Chen and Seeburg 1985). At the end of the run the gel is fixed, oven desiccated, and autoradiographed for 12–36 hr.

PCR-SSCP Analysis of *Mu*-Element Components of Ancient Maize

For enzymatic amplification, the agarose fragment containing ancient seed nucleic acids is melted and 1 µl of the agarose–nucleic acid suspension directly added to the PCR mixture (50 µl). Amplification is performed in a Perkin Elmer-Cetus Thermal Cycler (Perkin Elmer, Norwalk, Conn.) using 2.5 units *Thermus aquaticus* (Taq) polymerase, 0.3 µg each of the oligonucleotide primers described in the text (see Table 2), and 200 µM each of dATP, dCTP, dGTP, dTTP, resuspended in the reaction medium recommended by the polymerase manufacturer. Cycles (30) are set as follows: 94°C, 1 min (denaturation); 55°C, 20 sec (annealing); 72°C, 20 sec (elongation). Amplified DNA is fractionated on a $10 \times 7 \times 0.075$ cm, 15% polyacrylamide minigel (Midget Electrophoresis Unit, LKB, Bromma, Sweden), the gel cut up to isolate the relevant (90-bp) band, and the DNA eluted (by crushing and soaking), ethanol precipitated, and resuspended in 50 µl sterile distilled water (Sambrook et al. 1989). One tenth of the suspension is further amplified using cold dGTP, dTTP, and dCTP and $\alpha(^{32}P)$dATP. The radioactive DNA is passed through a G-50 DNA-grade Sephadex resin (Pharmacia, Uppsala, Sweden) to eliminate un-incorporated nucleotides, then desiccated using a Speed-Vac (Savant-Instr., Farmingdale, N.Y.) concentrator, and finally resuspended in sterile distilled water at a specific activity of about 5×10^5 cpm/µl. Subsequently 3 µl of radio-active DNA are added to 2 µl of a loading solution (95% formamide, 20 mM EDTA, 0.05% bromophenol blue, 0.05% xylene cyanol). The whole mixture is heated at 98°C for 3–5 min to denature the DNA, briefly centrifuged, and kept on ice until loading. The control sample (dsDNA) is obtained by mixing 2 µl radioactive DNA with 2 µl of sterile distilled water and 1 µl of loading solution (15% ficoll 400, 0.25% bromophenol blue, 0.25% xylene cyanol). Again using the Sequi-Gen apparatus, samples are run on a $50 \times 21 \times 0.025$ cm 8% polyacrylamide gel with 2% crosslinking containing 10% glycerol. The gel is kept in a coldroom (4°C). Electric parameters are set as follows: 15 W constant power for 4 hr followed by 12W for 9 hr. After the run the gel is fixed, oven desiccated, and autoradiographed for 2–5 hr at room temperature.

Sequence Analysis of PCR-amplified DNA

PCR-amplified DNA is fractionated on 2.5% low-temperature agarose or 15% polyacrylamide as indicated above, and the relevent amplification band purified. About 100–300 ng amplified DNA are phosphorylated using 3 units polynucleotide kinase, then repaired using 2 units T_4 DNA polymerase, and eventually cloned into plasmid pUC13 which previously has been digested into the *Sam*I site of the polylinker and dephosphorylated. Individual clones are sequenced using M13 primers and the dideoxy chain termination method adapted to supercoiled plasmid DNA sequencing (Chen and Seeburg 1985).

Precautions to Minimize DNA Carryover

To minimize carryover in amplification experiments the following precautions should be taken: (i) positive-displacement pipetting devices must be employed and reagents stored separately in small aliquots; (ii) gel electrophoresis equipment (gel tank and tray) used to fractionate nucleic acids isolated from ancient specimens should not have been used previously for modern plant nucleic acids; (iii) proper control extractions and amplifications reactions should be performed with DNA preparations not expected to carry the target sequences, e.g., with DNA purified from soil; (iv) no purification and/or enzymatic amplification of modern plant DNA should be performed in the laboratory where ancient samples are manipulated.

7. Molecular Analysis of an Ancient Maize Sample

7.1 DNA Isolation and Sequence Analysis

In this section we will demonstrate the practical application of some of the general guidelines given above by describing the molecular characterization of a sample of pre-Columbian maize. This investigation was part of an ongoing research project on maize origin and domestication.

Nine well-preserved maize ears were discovered by Giancarlo Ligabue (Centro Ricerche Studi Ligabue, Venice) in a Huari burial site (Peru coast). The maize ears were tentatively classified as proto-Pagaladroga or proto-Ancashino (R. Sevilla, pers. comm.) Maize kernels were submitted to radiocarbon age determination and found to be 980 ± 95 years old (reference year 1950). A preliminary examination by scanning electron microscope (SEM) of mechanically crushed seeds showed a well-preserved endosperm with starch granules and protein bodies. No evidence of bacterial or fungal contamination emerged from the SEM survey.

Nucleic acids were phenol extracted from single kernels. Fractionation of nucleic acids by gel electrophoresis on 2.5% agarose showed that the kernels from the Huari site contained low molecular weight nucleic acids (Fig. 1), a result which was further confirmed by terminal labeling of the nucleic acid fragments by polynucleotide kinase and radioactive ATP and fractionation on 8% polyacrylamide using a DNA sequencer (see Methods).

Regarding the precise nature of the nucleic acid fractions extracted from the kernels, a series of tests based on evaluation of RNase sensitivity, base composition, sequence homology, and antigenic properties clearly demonstrated that most ($> 90\%$) of the nucleic acid was rRNA (Venanzi and Rollo 1990; Rollo et al. 1991).

A further series of tests was performed to verify both the presence and the state of preservation of DNA in the maize specimen. Nucleic acid fractions isolated from single kernels were purified by fractionation on low-melting agarose to eliminate the brown contaminants, then submitted to enzymatic amplification using oligonucleotide systems designed to bind to short tracts of the mitochondrial cytochrome c oxidase subunit 1 gene (Cox 1) from fertile maize or to a highly repeated sequence (H2a) present in heterochromatin-rich maize. Details of the oligonucleotide systems are reported in Table 2. A parallel investigation performed with DNA isolated from modern maize, other

1 2 3 4 5 6 7 8 9 10 11 12 13 14 15 16 17 18 19

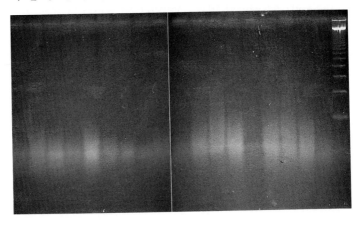

FIGURE 1. Gel electrophoresis of nucleic acids present in the Huari maize sample. Lanes 1–18, nucleic acid fractions isolated from single kernels of Huari maize; lane 19, molecular size marker (Bethesda Res. Labs "123-bp ladder"). Nucleic acid preparations were fractionated on 2.5% agarose containing ethidium bromide and the gel photographed under UV light.

TABLE 2. Inverse correlation between length of amplification system and amplification efficiency in Huari maize DNA

Length of PCR system (nucleotides)	Source of sequence	PCR efficiency (fraction of 10 preparations of ancient nucleic acid producing an amplification signal)
160	nuclear (H2a "long")	0/10
150	mitochondrial (Cox 1 "long")	0/10
130	mitochondrial (Cox 1 "short")	2–3/10
100	nuclear (H2a "short")	8–10/10
90	nuclear (*Mu*-terminal)	8–10/10

H2a "long":   GGCCACACAACCCCCATTTT, ACCCCATATGTTTCCTG
H2 "short":   TCAATAACATCGAAGGCTAAC, ACCCCACATATGTTTCCTG
Cox 1 "long": ACTTTTGCACCGAAGAAAT, ACGGACGGAATCACCGGGGA
Cox 1 "short": CATAAGTAATCCAATTTCCG, ACGGACGGAATCACCGGGGA
Mu-terminal: TGTTTACGCAGCCCCAAGTGCTGTC, GGCGTTGGCGTTGGCTTCTCTGTTT

graminaceous plants, filamentous fungi, and bacteria showed that the (Cox 1) amplification system was specific for maize DNA (Rollo et al. 1989). Such an amplification system can thus be applied to advantage to reveal the presence of endogenous genetic material in ancient maize remains.

The results of the amplification experiments using Cox 1, H2a, and other PCR systems are reported in Table 2. As shown in the table, Huari maize DNA displays a strong inverse correlation between efficiency of the amplification reaction and length of the target sequence, thus meeting the criterium of authenticity for ancient DNA stated by Pääbo et al. (1989). It seems of interest to point out that although this phenomenon has been attributed mainly to crosslinks in the template DNA, in the case of our maize sample the inverse correlation between amplification efficiency and length of the PCR target sequence appears to be a strict function of the length of the template DNA.

To further confirm the authenticity of the DNA sequences retrieved from the ancient seeds, the PCR products of the Cox 1 (short) system were purified by electrophoresis on low-melting agarose, submitted to restriction enzyme digestion, and fractionated on 20% polycrylamide. The results (Fig. 2) show that the amplified DNA is split into the fragments expected from the nucleotide sequence of modern maize DNA. Thus this first set of experiments clearly demonstrates that the kernels of the Huari maize contain DNA of endogenous origin. However, as stressed above, though PCR systems based on the enzymatic amplification of mtDNA sequences can be used to distinguish between

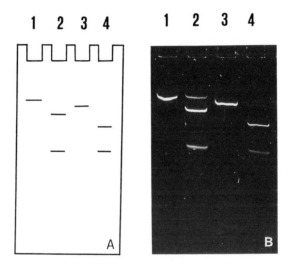

FIGURE 2. Comparison of the restriction patterns of modern and pre-Columbian (Huari) maize mtDNA (from nucleotide − 160 to nucleotide − 30 of the cytochrome *c* oxidase subunit 1 gene). (*A*) Computer-simulated restriction pattern of modern maize DNA. (*B*) Restriction digest of PCR-amplified Huari DNA. Lane 1, untreated control; lane 2, *Alu*I; lane 3, *Hae*III; lane 4, *Alu*I + *Hae*III. PCR-amplified DNA was fractionated on 2.5% low-melting agarose and the relevant band purified and submitted to enzymatic digestion. Treated DNA and untreated controls were fractionated on 20% polyacrylamide. The computer-simulated restriction pattern was obtained using the DNA Inspector II program (Textco, West Lebanon, N.H.) as previously reported (Rollo 1989). The uppermost band in lane 2 of the UV picture is due to undigested DNA.

DNA endogenous to the kernels and contaminant DNA, they are not genetically informative as they do not show polymorphism among different ancient
maize samples or between ancient and modern maize. For this reason, in
subsequent experiments only nuclear sequences were investigated, in particular
the distribution of *Mu*-element components in pre-Columbian maize.

All characterized *Mu* elements contain similar terminal inverted repeats of
about 200 bp length, yet the internal sequences of these elements may be
completely unrelated. It is known that the internal portion of the *Mu1* element
is represented in about one to three copies per haploid genome in every non-
Mutator stock and about 20 times in the *Mutator* stock. The terminal sequence
is more highly represented, with about 40 copies in all non-*Mutator* stocks
and about 60–70 copies in the *Mutator* genome. In addition, though *Mu*
elements are spread throughout the *Zea* genome they are absent in *Tripsacum*,
the genus most closely related to *Zea* (Talbert and Chandler 1988; Talbert
et al. 1989).

Due to the fact that terminal inverted repeats belonging to different *Mu*
elements show a high degree of homology, a PCR system designed to amplify
a DNA sequence located on these repeats is not expected to be specific for
any particular *Mu*-type element, but rather to give rise to a complex mixture
of amplified sequences deriving from the right or the left terminal repeat of
different elements. A PCR system of this sort is shown in Figure 3A, B. Amplified DNA can be cloned in a suitable vector and the nucleotide sequence of
individual clones determined. Results referring to the analysis of six individual
clones derived from the amplification of DNA isolated from a single Huari
maize kernel are shown in Figure 4. This type of analysis gives an indication
of both sequence variation and relative copy number of the different *Mu*-
element components in the ancient sample, by showing a relative abundance
of the *Mu4L*-element component. Recently, the analysis of very large sets
(> 800 in total) of recombinant clones, obtained from four independent amplification experiments using single-kernel DNA preparations, has allowed us to
confirm the relative abundance of the *Mu4L* component, and to determine
the presence of variant *Mu4*- and *Mu8*-element components in the Huari
maize DNA (Rollo et al., unpubl. results).

7.2 PCR-SSCP Characterization of
Mu-Elements Components

While an experimental approach such as the sequence analysis described
above of individual clones of amplified DNA can be fruitfully employed to
assess the different *Mu*-element components in DNA isolated from a single
kernel, or from a limited number of kernels, this procedure is far too laborious
and time consuming to be applied to the screening of the kind of large kernel
sample required for a population study. For this reason we have turned our
attention to alternative approaches.

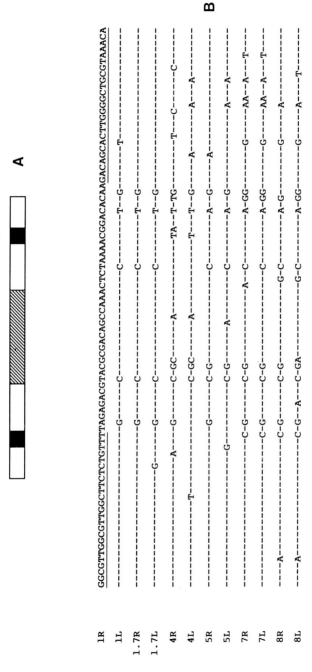

FIGURE 3. (*A*) The basic structure of a *Mu* transposable element showing the terminal inverted repeats (empty boxes) conserved within the *Mu* family and the internal portion (hatched) known to vary from element to element. The solid boxes within the repeat regions indicate the PCR target sites. The relative dimensions of the different *Mu*-element components are not maintained in the schematic. (*B*) Alignment of the nucleotide sequences of modern *Mu* terminal inverted repeats. Dashes indicate sequence identity. Underlining indicates the binding sites of the oligonucleotides used to amplify the selected portion (90-bp) of DNA.

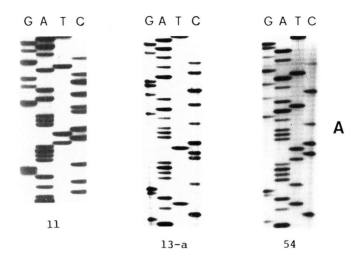

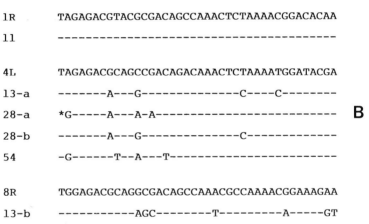

1R	TAGAGACGTACGCGACAGCCAAACTCTAAAACGGACACAA
11	--

4L	TAGAGACGCAGCCGACAGACAAACTCTAAAATGGATACGA
13-a	-------A---G--------------C----C--------
28-a	*G-----A---A-A--------------------------
28-b	-------A---G--------------C-------------
54	-G------T--A---T------------------------

8R	TGGAGACGCAGGCGACAGCCAAACGCCAAAACGGAAAGAA
13-b	-----------AGC--------T---------A-----GT

FIGURE 4. Sequence analysis of *Mu* terminal inverted repeats in Huari maize DNA. (*A*) Sequencing gels of clones 11, 13-a, and 54 of Huari maize DNA amplified using the PCR system shown in Figure 3. Arrows indicate the substitutions found when the Huari sequences are compared with those reported in the literature (modern standards). (*B*) Alignment of the nucleotide sequences of six clones of PCR-amplified Huari maize DNA with the corresponding sequence of known (modern) *Mu* elements. Dashes indicate identity with the standard sequences. The sequences of the two 25-mer oligonucleotide binding sites have been omitted.

A powerful technique for the rapid detection and characterization of allelic variants starting from PCR-amplified DNA is the mobility shift analysis of single-stranded DNA on neutral polyacrylamide gels (Hayashi 1991). DNA is PCR amplified and at the same time made radioactive either by use of labeled elongation primers or by addition of a labeled deoxynucleotide to the

PCR reagents. The amplified DNA is heat denatured and loaded onto a neutral low-crosslinking polyacrylamide gel, which is run at low power under carefully controlled temperature conditions (4–25°C, according to different protocols). Under these conditions, single-stranded DNAs of the same nucleotide length but of different nucleotide composition can be separated, probably due to differences in their predominant semi-stable conformations; hence the term *single-strand conformation polymorphism* (SSCP) to designate this feature of single-stranded DNAs.

SSCP analysis permits the separation of single-stranded fragments differing by as little as a single substitution. We have successfully applied this technique to resolve *Mu*-element component variants in our sample of Huari maize (Amici and Rollo 1992). The PCR-SSCP protocol developed in our laboratory is based on the *Mu*-specific amplification system. In principle, this system can produce 10 or more double-stranded 90-bp-long variants (left and right terminal inverted repeats of *Mu1*, *Mu4*, *Mu5*, *Mu7*, *Mu8*, and possibly other members of the *Mu* family not yet described). When run as single-strand DNA on a SSCP gel, these variants can be expected to give rise to a quite complicated banding pattern. To obtain a more simple picture we have taken advantage of the fact that the adenine composition of the two strands of the amplified sequence is significantly different (15–18 A residues in one strand, and 1–6 in the other). Thus, by using radioactive dATP as a substrate of the amplification reaction, "hot" and "cold" DNA strands can be obtained. Under appropriate exposure conditions the latter are expected to produce negligible banding, or no banding at all, on the autoradiograph of the SSCP gel. Results referring to the PCR-SSCP separation of *Mu*-element components of a Huari maize kernal are shown in Figure 5.

It may be worth noting that although SSCP analysis can reveal single mutations and thus potentially most incorporation errors as well, independent amplifications of the same preparation of Huari maize DNA have been shown to produce remarkably similar banding patterns. This phenomenon can be explained by the fact that presumably, due to the short length (90-bp) of the target sequence, DNA amplification initiates from a relatively large number of template molecules. As a consequence, polymerase errors are randomly distributed within the population and none of the erroneous sequences forms a subpopulation of significant size (Hayashi 1991). Preliminary experiments have shown that kernels from different ears of Huari maize produce similiar but not identical SSCP banding patterns.

8. Concluding Remarks

Ancient seeds can contain low-molecular weight rRNA and DNA of endogenous origin. DNA can be detected and characterized using proper PCR systems. The low degree of polymerization of the seed DNA and the mode of plant evolution impose a number of constraints on the recovery of genetic

1   2

FIGURE 5. PCR-SSCP banding pattern of *Mu* terminal inverted repeats in the DNA isolated from a single kernel of Huari maize: autoradiograph of an SSCP gel loaded with PCR-amplified DNA before (lane 1) and after (lane 2) thermic denaturation.

information. In general, polymorphism based on copy number, or on copy number and sequence variation, of a nuclear DNA tract appears to be better suited for diachronic genetic investigations of ancient seeds than polymorphism based on sequence variation alone. Full exploitation of the genetic information contained in ancient seeds also requires the availability of proper research tools for the more extensive investigation of populations. These tools now seem to be available.

Acknowledgments. The authors wish to thank Giancarlo Ligabue, Centro Ricerche Studi Ligabue, Venice; Lanfredo Castelletti, Civico Museo "Paolo Giovio," Como; and Silvio Curto and Enrichetta Leospo, Soprintendenza alle Antichità Egizie, Turin, for providing ancient seed samples. This research was supported by funding from the Ministero dell'Università e della Ricerca Scientifica assigned to F.R.

References

Amici A, Rollo F (1992) PCR-SSCP characterization of *Mu* element components in a Huari (X Century) maize sample. Ancient DNA Newsletter 1: in press

Chargaff E (1950) Chemical specificity of nucleic acids and mechanism of their enzymatic degradation. Experientia 6:201–240

Cheah KSE, Osborne DJ (1978) DNA lesions occur with loss of viability in embryos of ageing rye seeds. Nature 217:593–599

Chen EJ, Seeburg PH (1985) Supercoil sequencing: a fast simple method for sequencing plasmid DNA. DNA 4:165–170

Doebley J, Goodman MM, Stuber CW (1984) Isozyme variation in *Zea* (gramineae). Syst Bot 9:203–218

Doebley J, Renfroe W, Blanton A (1987) Restriction site variation in the *Zea* chloroplast genome. Genetics 117:139–147

Hayashi K (1991) PCR-SSCP: a simple and sensitive method for detection of mutations in the genomic DNA. PCR Methods and Applications 1:34–38

Higuchi RG, Wrischnik LA, Oakes E, George M, Tong B, Wilson AC (1987) Mitochondrial DNA of the extinct quagga: relatedness and extent of postmortem change. J Mol Evol 25:283–287

Meyer FG (1980) Carbonized food plants of Pompeii, Herculaneum, and the Villa at Torre Annunziata. Econ Bot 34:401–437

Ogram A, Sayeler GS, Barkay T (1988) DNA extraction and purification from sediments. J Microbiol Methods 7:57–66

Pääbo S (1985) Preservation of DNA in ancient Egyptian mummies. J Archaeol Sci 12:411–417

Pääbo S (1989) Ancient DNA: extraction, characterization, molecular cloning and enzymatic amplification. Proc Nat Acad Sci USA 86:6196–6200

Pääbo S, Higuchi RG, Wilson AC (1989) Ancient DNA and the polymerase chain reaction. J Biol Chem 264:9709–9712

Palmer JD, Herbon LJ (1988) Plant mitochondrial DNA evolves rapidly in structure, but slowly in sequence. J Mol Evol 28:87–97

Rivin CJ, Cullis CA, Walbot V (1986) Evaluating quantitative variation in the genome of *Zea mays*. Genetics 113:1009–1019

Rogers SO, Bendich AJ (1985) Extraction of DNA from milligram amounts of fresh, herbarium and mummified plant tissues. Plant Mol Biol 5:69–76

Rollo F (1985) Characterization by molecular hybridization of RNA fragments isolated from ancient (1400 B.C.) seeds. Theor Appl Genet 41:330–333

Rollo F (1989) Comparison of maize and teosinte DNA analysing actual and computer-simulated gel electrophoresis patterns of enzymatically amplified DNA. J Genet Breeding 43:179–184

Rollo F, La Marca A, Amici A (1987) Nucleic acids in mummified plant seeds: screening of twelve specimens by gel-electrophoresis, molecular hybridization and DNA cloning. Theor Appl Genet 73:501–505

Rollo F, Amici A, Salvi R, Garbuglia AR (1988) Small but faithful pieces of ancient DNA. Nature 335:774

Rollo F, Amici A, Salvi R, Garbuglia AR (1989) Characterization of a mitochondrial and nuclear maize DNA sequence by polymerase chain reaction: potential for use as molecular probes. J Genet Breeding 43:91–98

Rollo F, Venanzi FM, Amici A (1991) Nucleic acids in mummified plant seeds: bio-chemistry and molecular genetics of pre-Columbian maize. Genet Res 58:193–201

Sambrook J, Fritsch EF, Maniatis T (1989) *Molecular Cloning: A Laboratory Manual* 2nd ed. Cold Spring Harbor, N.Y.: CSHL Press

Schwarz-Sommer Z, Gierl A, Cuypers H, Peterson PA, Saendler H (1985) Plant trans-posable elements generate the DNA sequence diversity needed in evolution. EMBO J 3:591–597

Shewry PR, Kirkman MA, Burgess SR, Festenstein GN, Miflin BJ (1982) A comparison of the protein and amino acid composition of old and recent barley grain. New Phytol 90:455–466

Stollar BD (1986) Antibodies to DNA. In: *CRC Critical Reviews in Biochemistry 20*. Boca Raton, Fla.: CRC Press, pp. 1–36

Talbert LE, Chandler VL (1988) Characterization of a highly conserved sequence related to *Mutator* elements in maize. Mol Biol Evol 5:519–529

Talbert LE, Patterson GI, Chandler VL (1989) *Mu* transposable elements are struc-turally diverse and distributed throughout the genus *Zea*. J Mol Evol 29:28–39

Toole VK (1986) Ancient seeds: seed longevity. J Seed Technol 10:1–23

Venanzi FM, Rollo F (1990) Mummy RNA lasts longer. Nature 343:25–26

Zohary D, Hopf M (eds) (1988) *Domestication of Plants in the Old World* Oxford: Oxford University Press

Zurawski G, Clegg MT, Brown HD (1984) The nature of nucleotide sequence divergence between barley and maize chloroplast DNA. Genetics 106:735–749

Fossil Samples
17 DNA from Plant Compression Fossils

Edward M. Golenberg

1. Introduction

1.1 Fossil DNA in Evolutionary Studies

Evolutionary biology is a historical science. It seeks to understand the series of events that have led from a prior state or condition to a new, and hence derived, state or condition. In this quest, emphasis may be placed either on the reconstruction of the sequential order of events and the proper identification of intermediate states, or on the elucidation of the processes that led up to these events. In either approach, whether it is the pattern of events itself or the biological process that led to the pattern that is of interest, we need to assume the existence of a prior state, an origin where the evolutionary journey began and where the scientific journey must begin as well.

This prerequisite initial state makes every naturally occurring evolutionary lineage an experiment without replicate. Nonetheless, unlike patterns of human social and political history, many evolutionary patterns will be shaped by similar biological or stochastic processes so that the study of independent evolutionary histories can elucidate important common features. However, to fully exploit the information contained in these histories, it is necessary to have an understanding of their initial states. Only with this knowledge at hand can the rate and direction of change be determined.

Much of the theory of evolutionary biology has centered on the explanation of genetic variation within, and among natural populations of organisms. What maintains variation at a given gene locus? What determines the rate and trajectory of change of the genetic composition of a population and, by extension, of a species? How do physiological constraints, environmental stresses, reproductive and social behaviors contribute to the observed genetic state of the population under study? How important is the contribution of these biological factors compared to that of the stochastic factors that result from the transmission of genes between generations? In addition to these questions, the investigation of variation on the nucleotide sequence level over the past ten years has resurrected mutation as an interesting factor in shaping histories of genetic change. Similarly, evolution on the DNA/RNA level independent of or only secondarily affecting phenotype has become an active focus for investigation.

The processes that determine the genesis of an allele and its subsequent fate in a population or species fall under the conceptional framework of microevolution. In contrast, the study of macroevolution covers the genesis and fate of species through geological time periods. Since in the fossil record species differentiation is most often recognized by morphological change, macroevolutionary models are predominantly concerned with explaining the patterns of this morphological change. The relevance of microevolutionary processes to macroevolutionary patterns has been a source of controversy recently in evolutionary biology. Do the forces that determine allelic change have any bearing on morphological change and speciation, or does the isolation of gene pools associated with speciation simply permit the fine tuning or anagenesis of the newly isolated gene pools?

Superficially, work on ancient DNA, whether from museum skins (Higuchi et al. 1984; Thomas et al. 1990), exceptionally preserved soft tissue (Pääbo 1985a; Lawlor et al. 1991), or bones (Hagelberg et al. 1989, 1991; Hänni et al. 1990; Hagelberg and Clegg 1991; Horai et al. 1991; Hummel and Herrmann 1991), may appear to be indistinguishable from work on fossil DNA. Many of the technical aspects relating to endogenous damage to the DNA, extraction, and amplification are indeed shared. However, the time scales involved dictate the types of biologically relevant questions that can be addressed. Ancient DNA may be used to reconstruct proximal histories of species and populations. As a result, such material will provide fascinating windows to microevolutionary processes such as gene flow, domestication, population differentiation, inbreeding, and natural and artificial selection. Fossil DNA, by contrast, will probably not lend itself readily to population level studies, both because of the limited availability and variable quality of material at a given site and because of the problem of assessing whether samples are truly contemporaneous. Instead, the analysis of fossil DNA must be viewed in the context set up above. First, studies involving the extraction, sequencing, and verification of fossil DNA have demonstrated the existence of material that can be useful to both the paleontologist and the evolutionary geneticist. This opens the possibility for coordinated studies of macro- and microevolutionary patterns that will directly address the relationship between patterns and processes of morphological change on the one hand and genetic change on the other. Second, molecular evolutionary studies attempt to reconstruct relationships between contemporaneous taxa by inferring ancestral states and the distances between them. These inferences can be treated as testable hypotheses for which fossil DNA can provide experimental material. Third, fossil DNA sequences may provide a tool for the paleontologist to decouple the identification of species from morphology and, thereby, to address the problem of cladogenesis versus anagenesis. Fourth, expected patterns created by long-term processes such as nucleotide substitution or concerted evolution may be confirmed, depending on the future availability of lineages, by sampling over time.

1.2 DNA Preservation in Fossils

Eglinton and Logan (1991) have suggested that nucleic acids are among the molecules most susceptible to destruction over time. Intimate association with more resistant molecules may sequester more susceptible molecules from either chemical or biotic attack, thereby improving their preservation (Curry et al. 1991). Based upon this assertion, it would be reasonable to expect that a high degree of preservation of cellular and subcellular structures would be a necessary, though not sufficient, condition for the presence of preserved DNA or RNA. It should be noted that many structures may be visibly maintained through replacement of the original biomolecules by mineralization; in these cases, structure preservation would not necessarily reflect biomolecular preservation. However, this precaution notwithstanding, the ample reports of exceptionally fine preservation of cellular and subcellular structures do suggest that DNA-bearing fossils of a variety of ages may be found.

Perhaps the earliest report of nucleated fossil cells is that of Stevens (1912) regarding a Cretaceous palm; many subsequent reports have appeared in the literature. Schopf (1968) presented evidence of the oldest nucleated cells among individual Precambrian microfossils and suggested that some of these are a mitotic series. These fossils serve to demonstrate the degree of preservation attainable, but are not necessarily sources of nucleic acids. There are a number of reports on considerably younger plant fossils showing well-defined subcellular preservation. Darrah (1938) reported nuclei and chromosomes found in fossil *Selaginella*. Similarly, Baxter (1950) reported well-defined nuclei and chromatin granules in a Pennsylvanian calimitean species. Millay and Eggert (1974) also described nuclei in pollen from a Pennsylvanian seed fern, and Mamay (1957) reported nucleated Pennsylvanian coenopterid spore cells. Brack-Hanes and Vaughn (1978) reported the presence of exceptionally well-preserved nuclei in Pennsylvanian lycopod microgametophytes. They identified nucleoli in interphase cells, and mitotic cells in prophase, metaphase, and anaphase. Gould (1971) reported nuclei in Triassic cycad fossils and suggested that the organelles, while chemically altered, are organic and "not mineralogical segregations." Vishnu-Mittre (1967) published very distinctive micrographs of nuclei and chromosomes, including cells appearing to be in anaphase, from a Jurassic pteridophyte.

In addition to the evidence of preserved nuclei, there are several reports of well-preserved fossil chloroplasts and mitochondria. As with the publications mentioned above, micrographs in these reports vary greatly in quality and, therefore, in how compelling the data are. Baschnagel (1966) reported the presence of chloroplast remains in Devonian fossil algae. Bradley (1962) published a striking micrograph of an Eocene *Spirogyra* germling cell with a well-discernible spiral chloroplast. Similarly, Niklas and his colleagues published micrographs of chloroplasts demonstrating discernible stroma structures from Miocene angiosperm compression leaf fossils (Niklas 1983; Niklas et al. 1978, 1985). They also reported the presence of mitochondria in these fossils (Niklas et al. 1978, 1985).

These reports cannot be taken as evidence of the presence of fossil DNA. They do, however, suggest that an active, vigorous search for fossil DNA of a variety of ages is warranted. Unfortunately, it is not clear from prior work on ancient or fossil samples what the most suitable conditions are for preservation of DNA. Eglinton and Logan (1991) have recently reviewed what is known about preservation of biological macromolecules. Microbial activity is probably the predominant cause of degradation of organic material in general, and of nucleic acids in particular. The presence of internal proteases and nucleases will also contribute to the rapid degradation of dead organic material. The consistent pattern of greater decay of animal remains compared to plant remains is a reflection of these factors. Plant compression fossils, particularly leaf compression fossils, are less likely to have been senescent at the time of deposition, would not be expected to have had a significant internal or epiphytic bacterial fauna, and would be composed of a large percentage of microbe resistant molecules such as lignin and cutin. By contrast, animal compression fossils would more likely be derived from tissue that was dead at the time of deposition, had a significant internal microbiotic fauna contributing to its decomposition, and had a high protein content.

The nonbiotic conditions which may contribute to prolonged preservation are less consistently characterized. Pääbo and his colleagues (Pääbo 1985b, 1989; Pääbo and Wilson 1991) have suggested that rapid tissue dehydration is necessary for enhanced preservation of DNA, due to the susceptibility of the molecule to hydrolysis. This was inferred from studies of spontaneous depurination (Lindahl and Nyberg 1972) and from the suitability of various tissues for DNA analysis, as experimentally determined (Pääbo 1989). Niklas (1983) also suggested that rapid desiccation may contribute to the preferential preservation of some organelles in plant tissue, although for reasons related to properties of senescence in leaves. Perhaps relatedly, dehydration would be expected to reduce the activity of endogenous proteases and nucleases. However, a variety of DNA-containing ancient and fossil tissues have been investigated that were preserved fully hydrated (Golenberg et al. 1990; Hagelberg and Clegg 1991; Lawlor et al. 1991; Soltis et al. 1992). Initial studies of the base composition of at least one specimen do not suggest that the predominant damage is depurination (Golenberg 1991). The discrepancy between these findings and expectations of higher degradation may be due to an paucity of free metal ions which otherwise could generate hydroxyl radicals attacking the glycosyl and phosphodiester bonds (Eglinton and Logan 1991). Alternatively, hydration may reduce oxidative damage to the nitrogenous bases, in particular to the pyrimidines.

Similarly, at first sight, neutral conditions may be considered optimal for DNA preservation. Acidic conditions can depurinate the DNA, and alkaline conditions can cleave it at apurinic or apyrimidinic sites (Ausubel et al. 1991). In addition, alkaline conditions can denature double-stranded DNA and thereby eliminate the stability provided to the molecule by the hydrogen bonds between the two strands. Exposed bases would also be more susceptible to chemical modification and degradation. In light of these considerations, it is noteworthy that DNA has been extracted from soft tissue preserved in an acidic bog environment (Lawlor et al. 1991; cf. Hauswirth et al. Chapter III.2). Brack-Hanes and Vaughn (1978) suggest that chromosomal material, including nucleic acids is, in fact, better preserved under acidic conditions. Eglinton and Logan (1991) suggest that acidic conditions may initially inhibit microbial activity and, further, that humic acids can chelate metal ions, thereby contributing to prolonged preservation. Niklas et al. (1985) suggest that high concentrations of phenolics and tannic acids may contribute to fixing subcellular structure.

In light of the above discussion, some characteristics of deposition of fossil material that contribute to prolonged preservation can be identified. Rapid and directed deposition will minimize physical damage to the fossil material and reduce both externally induced (microbial) and internally induced (wound reactions, senescence, necrosis) degradation. Rapid sedimentation will also reduce exposure of the tissue to biotic degradation (Eglinton and Logan 1991; Yang 1993). Fine grain size and compaction along with volcanic interbeddings may prevent percolation and diffusion and, thereby, insulate the immediate environment of the fossil (Eglinton and Logan 1991; Yang 1993). Anoxic conditions, while not preventing microbial decomposition by themselves, may reduce oxidative damage. Yang (1993) has suggested that the presence of pyrites, carbonate minerals, and phosphate in sediments may also be indicative of sites bearing well-preserved fossils.

2. Extraction of Fossil DNA

2.1 Field Extraction of Fossils, Preliminary Processing, and Laboratory Precipitation of DNA

Based upon the above considerations, plant fossils that are found in lake deposits and are fully hydrated will probably be among the best, although not the only, sources for extraction of fossil DNA. It is likely that procedures used to extract DNA from leaf compression fossils at Clarkia, Idaho, and other localities (Golenberg et al. 1990; Golenberg 1991) will be adaptable to other sites. Equipment must be brought to the site to extract the fossil-bearing rock, expose the fossils, record the specimens, prepare them for DNA extraction, and transport them in a semi-stable state back to the laboratory for final DNA extraction and analysis. Table 1 lists the required equipment by task.

Because it is preferable to reduce the exposure of fossils to desiccation, oxygen, light, and external debris, processing of the material should proceed as rapidly as possible. Rock should be extracted from the deposit in a chunk large enough so that the probability of obtaining a complete specimen for identification is increased, yet small enough to be reasonably handled and split along the lamina. Exposed specimens should be identified immediately,

TABLE 1. List of equipment for field extraction

Task	Equipment
Excavation and exposure of fossils	Picks, shovels, knives
Recording of specimens	Camera, field book, labels, writing utensils, sample vials (for voucher specimen), forceps, alcohol
Processing of fossils	Fine spatulas or knives, glassine paper, mortars and pestles, dry ice, water, laboratory soap, wash bins, paper towels, 1.5-ml microfuge tubes, pipettors, disposable pipet tips, DNA extraction buffer
Transport of specimens to laboratory	Microfuge tube rack with lid, ice, cooler

photographed with identification numbers, and the information recorded in a field book. It may be useful, especially with rare or exotic specimens, to save a voucher sample of the cuticle for future reference. Voucher samples may be put in labeled vials with alcohol and stored on ice. Herbaria or museums may be likely repositories for such samples.

After initial processing, the compression fossil should be lifted or scraped from the rock and ground with a mortar and pestle. As with modern plant material, better results are obtained when the tissue is finely ground, presumably due to the disruption of the cell walls. The addition of small pellets of dry ice is helpful both physically in grinding the tissue and by reducing any damage that might arise from the heat given off in grinding. The ground tissue should be transferred to 1.5-ml microfuge tubes (glassine paper is helpful in the transfer). The ground material should fill no more than a third of the tube to allow for the addition of extraction buffer and chloroform/isoamyl alcohol. If sufficient material is available, the sample should be split among several different tubes. This allows for consistency checks on the extraction, and obviously provides for additional sources of material if a single sample is suspected of being contaminated. Extraction buffer (0.7 ml) is added, and the capped tubes are inverted or shaken to expose the grindate evenly to the buffer. The labeled tubes may be placed on ice or refrigerated at this stage until the extraction can be completed in the laboratory. Samples prepared in this manner have been stored several weeks before final extraction without any noticeable problems.

Negative control samples, both ones containing scraped clay without tissue and blank tubes, may also be collected at this time, simply as an exercise of "good field techniques." However, absence of DNA in levels visible under ethidium bromide staining in negative controls does not signify that DNA from fossil preparations is necessarily fossil in origin. Also, as some low level bacterial contamination must be expected considering the septic conditions of field work, presence of bacterial DNA in negative controls does not reflect on the authenticity of supposedly fossil DNA. As will be discussed shortly, actual verification must come from analysis of sequence data.

The choice of DNA extraction buffer is probably not critical; any standard plant DNA extraction buffer may be sufficient. Golenberg et al. (1990) and Soltis et al. (1992) used an extraction buffer modified from Rogers and Bendich (1985). The 2% buffer is made up as follows: 2% CTAB (w/v), 100 mM Tris (pH 8.00), 20 mM EDTA (pH 8.0), 1.4 M NaCl, and 1% PVP MW 40,000 (polyvinylpyrrolidone). An equally successful extraction buffer can be prepared substituting 2% SDS for 2% CTAB. In the event that only small quantities of DNA are expected, carrier molecules can be added to coprecipitate with the DNA. tRNA thus added will not interfere with later amplification or restriction of the DNA (Ausubel et al. 1991). Using oyster glycogen as carrier molecules instead will avoid the introduction of exogenous nucleic acids (Tracy 1981).

The final steps of DNA extraction can only be completed where controlled heating sources and centrifuges are available. Fossil grindates in extraction buffer are incubated at 65°C for approximately 1 hr. One-tenth volume of a solution of 5% sodium pyrophosphate and 5% phosphate glass is added to release DNA bound to clay particles (H. Wichman and D. Oliver, pers. comm.). The samples are incubated at 65°C for an additional 5 min, then centrifuged at $10,000 \times g$ at room temperature for 15 min to pellet debris. The supernatant is transferred to clean microfuge tubes and 0.7 ml (1 vol) chloroform:isoamyl alcohol (24:1 v:v) is added to each tube. The solutions are mixed by inverting the tubes, and the aqueous and organic layers are separated by a 5 min centrifugation ($10,000 \times g$). The aqueous (upper) layer is transferred to clean screw cap microfuge tubes. If glycogen is used, 1 μl (mg/ml) oyster glycogen solution is added at this point. One volume isopropanol is added to precipitate the DNA. The samples are mixed by inversion and stored at $-20°C$ overnight. The DNA is pelleted by centrifugation at $10,000 \times g$ for 20 min. The pellet can be washed twice in 70% ethanol, dried under vacuum, and reconstituted in TE buffer.

2.2 Secondary Purification of DNA

The presence of coprecipitating compounds such as polysaccharides, tannins, and humic acids is recognizable by a brown discoloration and a blue fluorescence under UV light. The blue fluorescing material contains sulfides (K. Spencer, pers. comm.) and migrates in agarose gels with a mobility comparable to DNA fragments 600 bp and smaller. These compounds can inhibit enzymatic amplification and therefore must be removed. If the concentration of fossil DNA allows, extractions can simply be diluted 1:50 or 1:100 before use in amplification reactions. Dilution will overcome the negative effects of the inhibiting compounds (Golenberg 1991). If large fragments of DNA are present, they may be separated from inhibiting compounds by agarose gel electrophoresis and subsequent extraction from the agarose.

Similar compounds have been removed from ancient bone-derived samples by spermine or spermidine DNA extraction (P. Persson, pers. comm.); that technique should be applicable to fossil DNA extractions as well. DNA is precipitated in the presence of 100 mM KCl and 10 mM spermine tetrachloride, pelleted and then washed in ethanol. Alternatively, a secondary CTAB extraction has proved successful at least in the removal of inhibiting polysaccharides (Murray and Thompson 1990; C. Giroux, pers. comm.). Vacuum-dried DNA is reconstituted in 5X TE buffer and 1/10 volume of 5% CTAB is added. After incubation on ice for 20 min, the precipitated CTAB and DNA are pelleted at 4°C by centrifugation at 3,000 × g. The supernatant is removed, and the CTAB is redissolved by addition of 70% ethanol, 200 mM sodium acetate (pH 5.5–6.0). This is then centrifuged at 10,000 × g for 20 min and the supernatant removed. This wash is repeated a total of three times, followed by subsequent 70% ethanol washes to remove the salts. Other methods, such as the use of glass beads or commercially prepared DNA-binding resins, also appear to remove some inhibitors, and may be applied in initial extraction procedures instead of alcohol precipitation.

3. DNA Isolation and Analysis

3.1 Criteria for Authenticity

It must be emphasized that extraction *per se* of nucleic acids from fossil material does not constitute demonstration of a fossil origin for this DNA. Neither can the behavior of particular DNA extracts in enzymatic amplification reactions—difficulty in amplification, size-specific amplification targets, negative results in negative extraction and amplification controls, or homogeneity of products—be used to support the authenticity of the extracted DNA. Recalcitrance to enzymatic amplification or restriction digests, homogeneity of products and true negative controls are all consistent with the possibility of outside contamination in extractions containing only inhibitory factors but no indigenous DNA. Similarly, apparent size-specific amplification products may be the result of chance primer incompatibility with specific contaminating DNA sequences. Unfortunately, much time must be invested in the analysis of such potential false positives.

Two unequivocal criteria have been suggested and explicitly or implicitly used to support identification of fossil and ancient DNA sequences (Higuchi et al. 1984; Golenberg et al. 1990; Hagelberg and Clegg 1991; Soltis et al. 1992). Unhappily, these criteria may be limited in their applicability to only a few systems. The first is that the derived sequences must be unique, that is, they must be demonstrably different from other known sequences. Singularity implies that the samples are not artifacts of contamination. This test requires implicitly that all possible sources of contamination, that is, all DNA that may come into contact with equipment in the field or in the laboratory, must be sequenced and added to the data base of known sequences. It should be obvious that this criterion may be inappropriate when multiple intraspecific samples, or ancient samples of modern species such as human or domesticated crops, are studied. The second criterion is that the sequences must be tested for phylogenetic authenticity, essentially based on similarity to sequences of modern species that are hypothesized to be closely related to the ancient or fossil species. This criterion will not be applicable to all studies either. Fossil remains may be mistakenly identified either due to an incomplete understanding of a particular group of organisms or due to the lack of diagnostic characters in the sample. Another possibility is that fossils may be of such antiquity that they represent taxa which are phylogenetically and temporally close to bifurcation events; that is, they may be closely related to shared common ancestors of modern lineages. If so, affinities based on sequence similarity would not be clear, due to the long time of divergence and the associated large variance of accumulated sequence differences. Essentially, the resolution of the position of short branches near the bifurcations of long branches on a phylogenetic tree may not be statistically feasible.

Because of these qualifications, the age of fossil and ancient DNA samples that can satisfactorily be tested according to the above two criteria may be limited to a certain window of time which will vary according to the groups of organisms under investigation. It is therefore incumbent upon investigators to exercise even greater care and skepticism than is normally demanded in good research programs.

3.2 PCR Amplification

The major considerations and problems of enzymatically amplifying specific DNA fragements from ancient and fossil DNA have been addressed extensively (Pääbo 1990; Golenberg 1991; Oste) and need not be discussed in great detail here. Let it suffice to summarize the main considerations relevant for, but not limited to, amplification of fossil DNA. The three major challenges are the presence of inhibitors, damaged and fragmented template DNA, and minimal source amounts of template DNA.

Inhibitors are most easily overcome by serial dilutions when these are possible. Secondary purification techniques as discussed above should also help reduce the concentration of inhibitors to below threshold levels. The use of excess levels of *Taq* polymerase enzyme (10 U or more) may overcome inhibitory effects as well; however, this may be a very expensive option, especially when a large number of samples needs to be processed. Excess enzyme may also contribute to an increased error rate in amplification (Eckert and Kunkel 1991).

The use of damaged and fragmented DNA as template DNA has two effects: the targeted fragment sizes may not exist in a given sample, and reconstruction of complete template strands through polymerization of partially overlapping fragments may increase the lag time of the reaction. The first problem can only be addressed by empirically testing a series of primers for each sample. The second effect is characterized by minimal amounts of amplified product after 30–40 cycles. Overlapping fragments serve as internal primers for single-stranded amplification. Cyclic amplifications of sequentially overlapping templates theoretically result in increasingly longer single-stranded fragments. However, until the complete fragment between the primer regions is reconstructed, there is no amplification of the desired product. Aliquots can be taken from such reactions and used as templates in a second series of 30–40 cycles to boost the product concentration. Appropriate negative controls should be set up for this procedure.

Minimal amounts of product are not likely a result of minimal amounts of template DNA as, theoretically, the exponential increase in concentration of product molecules can start out from a single template. Nevertheless, minimal template DNA can result in minimal product due to the random inefficiencies of the linked PCR component reactions. As discussed in Golenberg (1991), extending the denaturing and, perhaps, annealing times in the initial cycles may improve the overall efficiency of an amplification reaction.

The above caveats notwithstanding, amplification of fossil DNA can also be successfully attained under fairly standard conditions. For example, the amplification of part of the *rbc*L gene from the 17-million-year-old *Magnolia latahensis* (Golenberg et al. 1990) was completed in a 100-μl reaction with 50 pmoles of each end primer. The cycling conditions were denaturation for 1.5 min at 94°C, reannealing for 2 min at 38°C, and elongation for 3 min 72°C. The reaction solutions minus the enzyme were incubated at 94°C for 2 min prior to the addition of 2.5 U of *Taq* polymerase. Such an initial incubation both increases the likelihood of complete denaturation and prevents false priming when the solution is at lower temperatures. The two primers used in the amplification of the *Magnolia latahensis rbc*L fragment, Z234 and Z1024R, were designed by G. Zurawski and derived from the *Zea mays rbc*L sequence (Zurawski et al. 1984). For subsequent amplifications of fossil *rbc*L sequences, I have designed and successfully used sets of primers based on the *Liriodendron tulipifera rbc*L sequence (Golenberg et al. 1990). When amplifying from extant DNA, in which case larger fragments can be expected, I have designed and employed two additional primers, atpβ(c) (Golenberg et al. 1990) and rbcltts, which flank the entire *rbc*L region. All primers are listed in Table 2.

3.3 Sequencing Techniques

The inherent error rate of DNA polymerases guarantees that PCR products will be composed of a population of molecules differing by base substitutions. However, the low error rate of these enzymes and the randomness of errors along a DNA sequence will almost always result in the great majority of molecules having the correct nucleotide in a given position. Direct sequencing of PCR products takes advantage of this "truth by majority rule". Randomly produced errors at given bases will appear as distinctively weaker bands at

TABLE 2. *rbc*L Sequencing and amplification primers

Primer[a]	Sequence	Source of sequence[d]
Z234	CGTTACAAAGGACGATGCTACCACATCGAG	Zurawski et al. 1984
Z1053R	ATCATCACGTAGTAAATCAACAAAGCCTAAAGT	Zurawski et al. 1984
Ltrbcl 1	ATGTCACCACAAACAGAGACTAAAGC	Golenberg et al. 1990
Ltrbcl 209	GGACCGATGGACTTACCAGCCTTGATCG	Golenberg et al. 1990
Ltrbcl 478	GATAAATTGAACAAGTATGGTCGTCC	Golenberg et al. 1990
Ltrbcl 724	GCTACTGCAGGTACATGCGAAG	Golenberg et al. 1990
Ltrbcl 955	CGTATGTCTGGTGGAGATCATATTCACTC	Golenberg et al. 1990
Ltrbcl 1201	CAGTTCGGTGGAGGAACTTTAGG	Golenberg et al. 1990
Ltrbcl 1366R	CCTTCCATACCTCACAAGCAGCAGC	Golenberg et al. 1990
rbcltts[b]	GATTGGGCCGAGTTTAATTGC	Zurawski, Clegg 1987
atpβ(c)[c]	GTGGAAACCCCGGGACGAGAAGTAGT	Krebbers et al. 1982; Shinozaki et al. 1986

[a]The designations of primers are based on the species from which the sequence is derives (Z = *Zea mays*, Lt = *Liriodendron tulipifera*) and the position of the first nucleotide of the primer within that sequence (the start methionine codon is numbered 1–3). Reverse primers are made using the reverse complimentary sequence. For example, Ltrbcl 209R is CGATCAAGGCTGGTAAGTCCATCGGTCC.
[b]This sequence is based on the *rbc*L transcription termination sequence in *Nicotiana tabacum* 3′ to the protein encoding region. The sequence is highly conserved in comparison to homologous sequences in *Phaseolus albus*, *Pisum sativum*, and *Spinacia oleracea*.
[c]This primer is designed from a consensus sequence from *atp*β in the chloroplast genomes of *Nicotiana tabacum* and *Zea mays*. The sequence is a reverse complement from positions 41–16 (the start methionine codon is numbered 1–3).
[d]Papers from which original sequences were taken to design the primer.

particular positions along the gene sequence if at all. Cloning PCR fragments into plasmid vectors, the alternative method, randomly samples single molecules from the population, but numerous clones must be sequenced to assay the variance of the products. Thus, whenever possible, direct sequencing should be viewed as the preferred method. Even when multiple sequences are expected among the PCR products, such as when a heterozygous nuclear gene is assayed, identification of single állele sequences can be carried out (Clark 1990). Only when multi-copy genes are amplified would direct sequencing not be the method of choice because it may be impossible, in that case, to distinguish between error-produced and true indigenous variation.

Direct sequencing of PCR products entails either double-strand sequencing of a normal amplification reaction or sequencing of a single strand produced in a primary asymmetric or secondary single primer reaction (Kreitman and Landweber 1989; Ausubel et al. 1991). The sequencing protocols are identical to those for plasmid-produced DNA. In the case of fossil DNA, I prefer using a two-step approach as suggested by Kreitman and Landweber (1989). A double-strand amplification reaction is completed first. This first reaction provides a reservoir of synthetic double-stranded DNA, so that it is not necessary as often to go back to original fossil samples. An aliquot from this reaction is then used as template in a second reaction in which 20 pmoles of only one primer is used. The reaction conditions are essentially identical to those of the first reaction, although higher annealing temperatures may be used as the primer and target sequences are now 100% complementary. The single-stranded product is then purified using high salt precipitation (Kreitman and Landweber 1989) and then sequenced directly using internal primers such as those listed in Table 2.

Recent developments involving thermally stable polymerases and the concept of cycling to increase product yield have made sequencing of very small amounts of template DNA possible (Samols and Fuller 1991). The potential advantage of such amplification techniques for the analysis of fossil or ancient DNA is most obvious for the sequencing of fragments with one undetermined endpoint. As discussed above, amplification of fossil DNA may be unsuccessful if complete fragments of a targeted length are not available. By generating fragments of increasing length from one primer site, a sequence with no predetermined endpoint could be produced. This could, under the best of conditions, eliminate the costly process of generating a set of nested primers to determine what fragment of DNA is present for amplification.

4. Prospects for Fossil DNA Research

4.1 Systematics

Systematics has been the area of evolutionary biology that has been changed most radically by the use of recombinant DNA technology. The use of DNA sequence data, alone or in conjunction with morphological and behavioral characteristics, has had a rapid and profound effect on our ability to reconstruct the evolutionary histories of taxa and to classify them accordingly. The analysis of fossil DNA sequences will have its most immediate impact on the systematics of fossil species themselves. Fossil DNA will be used to identify and classify problematic specimens and to verify previous identifications and classifications of well-studied fossil groups. In addition, the placement of fossil taxa in phylogenetic trees based on extant taxa will serve as a test of phylogenetic hypotheses. Phylogenetic trees can be viewed as a set of hypotheses on the existence and characteristics of extinct taxa. Inclusion of fossil sequences in such trees will allow for improved evaluation of the comparative predictiveness of different phylogenetic or phenetic tree-generating algorithms, and of the reliability for phylogenetic inferences of the information contained in a given DNA sequence. Lastly, phylogenetic analyses of fossil DNA sequences will themselves produce evolutionary hypotheses regarding the evolutionary histories of extant lineages, such as whether certain clades are anagenic or reticulate. These hypotheses will in turn be testable by analysis of younger and older fossils within a clade.

4.2 DNA Diagenesis and Sequence Evolution

A major area of interest in the study of molecular evolution concerns the rate of accumulation of nucleotide substitutions (e.g. Gillespie 1986; Kimura 1987; Li and Tanimura 1987). While the accumulation of nucleotide substitutions should be monotonic, that is, more substitutions should occur over longer periods, it is not clear whether the rate of accumulation is constant or not. Under neutral theory, the rate of nucleotide substitution should be directly proportional to the mutation rate, but may be modified by constraints on the gene product (Kimura 1983). The mutation rate itself should not be affected by selection, population size, or speciation events. Therefore, rates of substitution should remain constant over time (Zuckerkandl and Pauling 1965; Kimura 1987). Alternatively, under models of selection, the rate of nucleotide substitution will be related to the product of mutation rate, selection coefficients, and population size. Both selection coefficients and population size will vary over time, and therefore, under a selection model, rates should not remain constant.

Estimated rates of the nucleotide substitution are commonly derived by estimating the number of nucleotide substitutions between two extant species and dividing that number by twice the time since divergence from the last shared ancestor (Fig. 1). Determination of the time since divergence is often

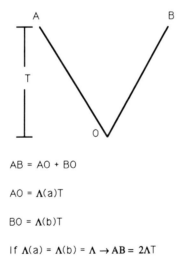

$$AB = AO + BO$$

$$AO = \Lambda(a)T$$

$$BO = \Lambda(b)T$$

$$\text{If } \Lambda(a) = \Lambda(b) = \Lambda \rightarrow AB = 2\Lambda T$$

FIGURE 1. Method of estimation of rates of nucleotide substitution via reference to a common ancestor. The number of nucleotide substitutions is represented by branch lengths of the phylogenetic tree. Rates (λ) of nucleotide substitution leading from the common ancestor O to each of the extant taxa A and B are assumed to be equal and constant, that is, $\lambda_A = \lambda_B$. The numbers of nucleotide substitutions between A and O and B and O are $\lambda_A T$ and $\lambda_B T$, respectively, where T is the time since divergence from the last common ancestor (to the present). Therefore, the expected number of substitutions between A and B is $\lambda_A T + \lambda_B T = 2\lambda T$. The rate λ can be estimated by dividing the number of nucleotide substitutions between A and B by 2T.

difficult, as it is based on fossil records, and can lead to disparate figures (e.g., Wolfe et al. 1987; Zurawski and Clegg 1987). An alternative method of detecting differences in rate is the relative rates test (Sarich and Wilson 1973; Wu and Li 1985). This test avoids direct dependence on the time of divergence and relies instead on a difference in substitutions between two test species and a third taxon (Fig. 2). Such a lest is effective in determining variations in rate among extant taxa, but it is not useful in estimating actual rates of

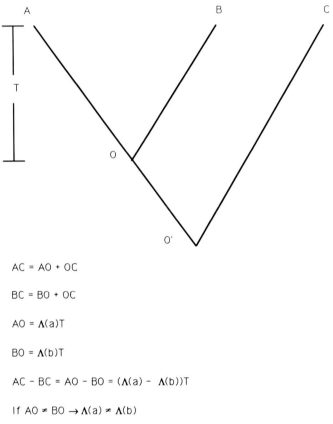

$$AC = AO + OC$$

$$BC = BO + OC$$

$$AO = \Lambda(a)T$$

$$BO = \Lambda(b)T$$

$$AC - BC = AO - BO = (\Lambda(a) - \Lambda(b))T$$

$$\text{If } AO \neq BO \rightarrow \Lambda(a) \neq \Lambda(b)$$

FIGURE 2. Relative rate test. The number of nucleotide substitutions is represented by branch lengths of the tree. The branch length between extant taxa A and C (AC) is equal to the sum of the branches AO and OC. Similarly, the branch length between extant taxa B and C is equal to the sum of BO and OC. The difference between branches AC and BC (AC − BC) is equal to the difference between branches AO and BO (AO − BO). Since AO, the number of nucleotide substitutions between A and the common ancestor O, is equal to $\lambda_A T$, where λ_A is the rate of nucleotide substitutions along the lineage leading to taxon A, and BO is equal to $\lambda_B T$, the difference AO − BO is euqal to $(\lambda_A T - \lambda_B T)$, where T is the time since divergence from the last common ancestor (to the present). If this difference is not equal to zero, then λ_A is not equal to λ_B.

substitution. In addition, both direct estimates of rates and tests of relative rates cannot detect nonhomogeneity of rates along a single lineage, since both methods look at rates averaged over long periods of times. Variations in rate resulting from speciation events or rapid morphological evolution, which may be associated with changes in population size or selection coefficients, would not be discernable.

Fossil DNA may be exploited both to estimate rates of nucleotide substitutions and to detect variations in rate within a lineage. By modifying the relative rates test to accommodate the different time periods along branches from a common ancestor to an extant species and to an extinct species, respectively, actual rates may be estimated (Fig. 3). Assuming that the rates are constant along each of the two branches—which may not always hold— rate differences between the branches may be viewed as the product of the substitution rate and the difference in time of divergence from a common ancestor between the extinct and extant species, or $\lambda(T - T')$. Because this difference in time is simply the time of deposition of the fossil before the present, independent information on the time of divergence is not necessary. Rates of nucleotide substitution may therefore be estimated from the difference in number of nucleotide substitutions compared to a third species, divided by the time of deposition. The assumption of uniformity of rates within a lineage could be tested whenever fossils of different age are available within the same lineage. Any detected variation in rate would violate assumptions of uniformity and would therefore weaken confidence in rate estimates; however, such variation could then be correlated with morphological and species distribution information to address the relative importance of non-neutral factors for sequence evolution.

Of course, a major problem that must be addressed in the study of fossil DNA must be the extent to which damaged template DNA may contribute to observed nucleotide differences. Simply put, we must determine how good the information is that we are getting from fossil DNA. Damage along a strand of DNA, such as AP sites or pyrimidine crosslinks, may be illegitimately repaired in polymerase reactions and thus result in higher apparent rates of evolution. Special concern must be given to systems where excess enzyme is added to achieve amplification. Under certain conditions, polymerases are able to read through lesions by incorporationing noncomplementary nucleotides. These completed but inauthentic copies of the original DNA strand would serve as the predominant template, and thus unambiguous but false product sequences would arise. Multiple independent samples of the same species must be surveyed to assess the extent of this potential problem. When not enough sample material is available, the distribution and types of substitutions must be analyzed to alert oneself to the possibility of such errors.

4.3 Evolutionary Model Testing

Recent theoretical consideration of the patterns of sequence evolution has suggested a variety of models for DNA sequence evolution, including neutral

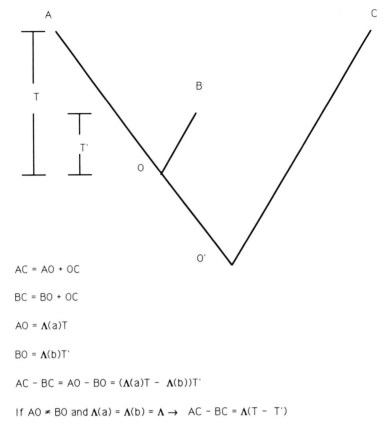

AC = AO + OC

BC = BO + OC

AO = Λ(a)T

BO = Λ(b)T'

AC − BC = AO − BO = (Λ(a)T − Λ(b))T'

If AO ≠ BO and Λ(a) = Λ(b) = Λ → AC − BC = Λ(T − T')

FIGURE 3. Method of estimation of rates of nucleotide substitution via reference to the time of fossil deposition. A and C are extant taxa. B is a fossil taxon within the clade leading from common ancestor O' to A. The branch lengths between taxa represent numbers of nucleotide substitutions and are the product of time and the rate of nucleotide substitution. As in the relative rates test, the branch length between extant taxa A and C (AC) is equal to the sum of branches AO and OC, and the branch length between fossil taxon B and extant taxon C is equal to the sum of BO and OC. The difference between branches AC and BC (AC − BC) is equal to the difference between branches AO and BO (AO − BO). AO is equal to λ_AT, where λ_A is the rate of nucleotide substitution along the lineage leading to taxon A, and T is the time since the divergence from ancestor O (to the present). BO is equal to λ_BT', where λ_B is the rate of nucleotide substitution along the lineage leading to B, and T' is the time from the divergence from ancestor O until the fossil was deposited. The difference AO − BO is equal to (λ_AT − λ_BT'). If the rates are equal, this difference is equal to λ(T − T'). The difference in times (T − T') is simply equal to the time since the fossil was deposited and is independent of the time of divergence from O. The rate of nucleotide substitution, λ, can therefore be estimated by dividing the difference in number of nucleotide substitutions between A and C and B and C by the time of fossil deposition.

nucleotide substitution, slipped-strand replication (Levinson and Gutman 1987), unequal crossing-over (Ohta 1983), gene conversion (Jeffreys 1979), exon shuffling (Miyata et al. 1980), and horizontal transfer (Syvanen 1986). In many cases, different mechanisms or collections of mechanisms would be expected to result in identical patterns; however, all evolutionary developments must consist of a sequence of subevents, and the intermediate states or conditions in such a sequence may vary with the mechanism of evolutionary change. Multiple sampling over time within a lineage may be helpful in differentiating between such mechanisms, by testing for the presence of hypothesized transition states. Some processes, such as neutral substitution, or even slipped-strand replication and exon shuffling, may produce gradual, incremental changes in the nucleotide sequence, whereas other processes, such as unequal crossing-over, gene conversion, or horizontal transfer, may produce distinctively discontinuous patterns of sequence evolution. Studies of the patterns of sequence change over time will allow finer appreciation of the potential variety of genetic and evolutionary processes that may occur in different genes over time.

5. Gene Product Analysis

The ability to amplify protein-encoding genes from fossils and then to clone them into expression vectors may open an avenue for analysis of not only genes but gene products. Judicious choice of genes for such analysis may allow us to address aspects of the evolution of protein-encoding genes in terms of the properties of the proteins themselves. Under well-designed experiments, it may be possible to include certain aspects of physiological ecology in the study of paleoecology. Where the pathways of secondary metabolites are well defined for a system, in vitro reconstruction may be attainable, opening up the possibility of the study and exploitation of ephemeral fossil compounds.

The ultimate assessment of the importance of fossil DNA research will not be a judgment of the technical finesse involved, nor will it depend on the discovery of "the oldest DNA". The value of the investigation of fossil DNA lies in the types of questions we can pose in light of this material, and the understanding we can gain from the information to be found in it. Naturally, for the next few years, active exploration and empirical development of techniques for fossil DNA recovery and analysis will be the focus of our studies. But beyond this initial phase if not sooner, fossil DNA research can and should play an active role in expanding our understanding of evolutionary theory.

Acknowledgment

This work was supported in part by a Wayne State University Research Grant and by an Undergraduate Biological Sciences Education Initiative Grant to Wayne State University from the Howard Hughes Medical Institute.

References

Ausubel FM, Brent R, Kingston R, Moore DD, Seidman JG, Smith JA, Struhl K (eds) (1991) *Current Protocols in Molecular Biology.* New York: Greene Publishing Associates and Wiley-Interscience

Baschnagel RA (1966) New fossil algae from the Middle Devonian of New York. Trans Amer Microsc Soc 85:297–302

Baxter RW (1950) *Peltastrobus reedae:* A new sphenopsid cone from the Pennsylvanian of Indiana. Bot Gaz 112:174–182

Brack-Hanes SD, Vaughn JC (1978) Evidence of paleozoic chromosomes from lycopod microgametophytes. Science 200:1383–1385

Bradely WH (1962) Chloroplast in *Spirogyra* from the Green River formation of Wyoming. Amer J Sci 260:455–459

Clark AG (1990) Inference of haplotypes from PCR-amplified samples of diploid populations. Mol Biol Evol 7:111–122

Curry GB, Cusack M, Walton D, Endo K, Clegg H, Abbott G, Armstrong H (1991) Biogeochemistry of brachiopod intracrystalline molecules. Phil Trans R Soc Lond B 333:359–366

Darrah WC (1938) A remarkable fossil Selanginella with preserved gametophyte. Botanical Museum Leaflets, Harvard University 6:113–136

Eckert KA, Kunkel TA (1991) DNA polymerase fidelity and the polymerase chain reaction. PCR 1:17–24

Eglinton G, Logan GA (1991) Molecular preservation. Phil Trans R Soc Lond B 333:315–328

Gillespie JH (1986) Variability of evolutionary rates of DNA. Genetics 113:1077–1091

Golenberg EM (1991) Amplification and analysis of Miocene plant fossil DNA. Phil Trans R Soc Lond B 333:419–427

Golenberg EM, Giannassi DE, Clegg MT, Smiley CJ, Durbin M, Henderson D, Zurawski G (1990) Chloroplast DNA sequence from a Miocene *Magnolia* species. Nature 344:656–658

Gould RE (1971) *Lyssoxylon grigsbyi,* a cycad trunk from the upper Triassic of Arizona and New Mexico. Amer J Bot 58:239–248

Hagelberg E, Clegg JB (1991) Isolation and characterization of DNA from archaeological bone. Proc R Soc Lond B 244:45–50

Hagelberg E, Sykes B, Hedges R (1989) Ancient bone DNA amplified. Nature 342:485

Hagelberg E, Bell LS, Allen T, Boyde A, Jones SJ, Clegg JB (1991) Analysis of ancient bone DNA: techniques and applications. Phil Trans R Soc Lond B 333:399–407

Hänni C, Laudet V, Sakka M, Begue A, Stehelin D (1990) Amplification of mitochondrial DNA fragments from ancient human teeth and bones. C r Acad Sci Paris, Ser III. 310:365–370

Higuchi R, Bowman M, Friedberger M, Ryder OA, Wilson AC (1984) DNA sequences from the quagga, an extinct member of the horse family. Nature 312:282–284

Horai S, Kondo R, Murayama K, Hayashi S, Koike H, Nakai N (1991) Phylogenetic affiliation of ancient and contemporary humans inferred from mitochondrial DNA. Phil Trans R Soc Lond B 333:409–417

Hummel S, Herrmann B (1991) Y-Chromosome-specific DNA amplified in ancient human bone. Naturwissenschaften 78:266–267

Jeffreys A (1979) DNA sequence variants in $^{G}\gamma$, $^{A}\gamma$-, δ- and β-globin genes of man. Cell 18:1–10

Kimura M (1983) The neutral theory of molecular evolution. In: Nei M, Koehn RK (eds) *Evolution of Genes and Proteins.* Sunderland, Mass.: Sinauer Associates.

Kimura M (1987) Molecular evolutionary clock and neutral theory. J Mol Evol 26:24–33

Krebbers ET, Larrinua IM, McIntosh L, Bogorad L (1982) The maize chloroplast genes for the β and ε subunits of the photosynthetic coupling factor CF_1 are fused. Nucl Acids Res 10:4985–5002

Kreitman M, Landweber LF (1989) A strategy for producing single-stranded DNA in the polymerase chain reaction: a direct method for genomic sequencing. Gene Anal Tech 6:84–88

Lawlor DA, Dickel CD, Hauswirth WW, Parham P (1991) Ancient *HLA* genes from 7,500-year-old archaeological remains. Nature 349:785–788

Levinson G, Gutman GA (1987) Slipped-strand mispairing: a major mechanism for DNA sequence evolution. Mol Biol Evol 4:203–221

Li W-H, Tanimura M (1987) The molecular clock runs more slowly in man than in apes and monkeys. Nature 326:93–96

Lindahl T, Nyberg B (1972) Rate of depurination of native deoxyribonucleic acid. Biochem 11:3610–3618

Mamay SH (1957) *Biscalithea,* a new genus of Pennsylvanian coenopterids, based on its fructification. Am J Bot 44:229–239

Millay MA, Eggert DA (1974) Microgametophyte development in the paleozoic seed fern family Callistophytaceae. Am J Bot 61:1067–1075

Miyata T, Yasunaga T, Yamawaki-Kataoka Y, Obata M, Honjo T (1980) Nucleotide sequence divergence of mouse immunoglobulin γ_1 and γ_{2b} chain genes and the hypothesis of intervening sequence-mediated domain transfer. Proc Natl Acad Sci USA 77:2143–2147

Murray MG, Thompson WF (1990) Rapid isolation of high-molecular-weight plant DNA. Nucl Acid Res 8:4321

Niklas KJ (1983) Organelle preservation and protoplast partitioning in fossil angiosperm leaf tissues. Am J Bot 70:543–548

Niklas KJ, Brown RM Jr., Santos R, Vian B (1978) Ultrastructure and cytochemistry of Miocene angiosperm leaf tissues. Proc Natl Acad Sci USA 75:3263–3267

Niklas KJ, Brown RM Jr., Santos R (1985) Ultrastructural states of preservation in Clarkia angiosperm leaf tissues: implications on modes of fossilization. In: Smiley CJ (ed) *Late Cenozoic History of the Pacific Northwest.* San Francisco: Pacific Division of the American Association for the Advancement of Science, pp. 143–160

Ohta T (1983) On the evolution of multigene families. Theor Pop Biol 23:216–240

Pääbo S (1985a) Molecular cloning of ancient Egyptian mummy DNA. Nature 314:644–645

Pääbo S (1985b) Preservation of DNA in ancient Egyptian mummies. J Archeol Sci 12:411–417

Pääbo S (1989) Ancient DNA: extraction, characterization, molecular cloning, and enzymatic amplification. Proc Natl Acad Sci USA 86:1939–1943

Pääbo S (1990) Amplifying ancient DNA. In: Innis UA, Gelfland DH, Sninsky JJ, White TJ (eds) *PCR Protocols: A Guide to Methods and Applications.* San Diego: Academic Press, pp. 159–166

Pääbo S, Wilson AC (1991) Miocene DNA sequences—a dream come true? Current Biology 15 46

Rogers SO, Bendich AJ (1985) Extraction of DNA from milligram amounts of fresh, herbarium and mummified plant tissues. Plant Mol Biol 5:69–76

Samols SB, Fuller CW (1991) Using cycled labeling reactions for cycle sequencing. Comments 18:23–25

Sarich VM, Wilson AC (1973) Generation time and genomic evolution in primates. Science 179:1144–1147

Schopf JW (1968) Microflora of the bitter springs formation, late Precambrian, central Australia. J Paleont 42:651–688

Shinozaki K, Ohme M, Tanaka M, Wakasugi T, Hayashida N, Matsubayashi T, Zaita N, Chunwongse J, Obokata J, Yamaguchi-Shinozaki K, Ohto C, Torazawa K, Meng BY, Sugita M, Deno H, Kamogashira T, Yamada K, Kusuda J, Takaiwa F, Kato A, Tohdoh N, Shimada H, Sugiura M (1986) The complete nucleotide sequence of the tobacco chloroplast genome: its gene organization and expression. EMBO J 5:2043–2049

Soltis PS, Soltis DE, Smiley CJ (1992) An rbcL sequence from a Miocene Taxodium (bald cypress). Proc Natl Acad Sci USA 89:449–451

Stevens NE (1912) A palm from the upper Cretaceous of New Jersey. Am J Sci 34:421–436

Syvanen M (1986) Cross-species gene transfer: a major factor in evolution? TIG 2:63–66

Thomas WK, Pääbo S, Villablanca FX, Wilson AC (1990) Spatial and temporal continuity of kangaroo rat populations shown by sequencing mitochondrial DNA from museum specimens. J Mol Evol 31:101–112

Tracy S (1981) Improved rapid methodology for the isolation of nucleic acids from agarose gels. Prep Biochem 11:251–268

Vishnu-Mittre (1967) Nuclei and chromosomes in a fossil fern. In: Darlington CD, Lewis KR (eds) Chromosomes Today. New York, Plenum pp. 250–251

Wolfe KH, Li W, Sharp P (1987) Rates of nucleotide substitution vary greatly among plant mitochondrial, chloroplast and nuclear DNAs. Proc Natl Acad Sci USA 84:9054–9058

Wu C, Li W (1985) Evidence for higher rates of nucleotide substitution in rodents than in man. Proc Natl Acad Sci USA 82:1741–1745

Yang H (1993) Miocene lake basin analysis and comparative taphonomy: Clarkia (Idaho, USA) and Shanwang (Shandong, PR China). Ph.D. thesis, University of Idaho

Zuckerkandl E, Pauling L (1985) Evolutionary divergence and convergence in proteins. In: Bryson V, Vogel H (eds) Evolving Genes and Proteins. New York: Academic Press

Zurawski G, Clegg MT, Brown AHD (1984) The nature of nucleotide sequence divergence between barley and maize chloroplast DNA. Genetics 106:735–749

Zurawski G, Clegg MT (1987) Evolution of higher plant chloroplast DNA-encoded genes: implications for structure-function and phylogenetic studies. Annu Rev Plant Physiol 38:391–418

Index